Contents

Preface vii

1 Coordinate Systems 1

1-1 Introduction 1
1-2 Cartesian Coordinates 2
1-3 Plane Polar Coordinates 4
1-4 Spherical Polar Coordinates 5
1-5 The Complex Plane 7

2 Functions and Graphs 11

2-1 Functions 11
2-2 Graphical Representation of Functions 12
2-3 Roots to Polynomial Equations 20

3 Logarithms 24

3-1 Introduction 24
3-2 General Properties of Logarithms 24
3-3 Common Logarithms 25
3-4 Natural Logarithms 27

4 Differential Calculus 30

4-1 Introduction 30
4-2 Functions of Single Variables 31
4-3 Functions of Several Variables—Partial Derivatives 35
4-4 The Total Differential 37
4-5 Derivative as a Ratio of Infinitesimally Small Changes 40
4-6 Geometric Properties of Derivatives 43
4-7 Constrained Maxima and Minima 47

5 Integral Calculus 54

5-1 Introduction 54
5-2 Integral as an Antiderivative 55
5-3 General Methods of Integration 55
5-4 Special Methods of Integration 57
5-5 The Integral as a Summation of Infinitesimally Small Elements 59
5-6 Line Integrals 61
5-7 Double and Triple Integrals 63

6 Differential Equations 69

6-1 Introduction 69
6-2 Linear Combinations 70
6-3 First-Order Differential Equations 71
6-4 Second-Order Differential Equations with Constant Coefficients 74
6-5 General Series Method of Solution 77
6-6 Special Polynomial Solutions to Differential Equations 79
6-7 Exact and Inexact Differentials 84
6-8 Integrating Factors 87
6-9 Partial Differential Equations 88

7 Infinite Series 94

7-1 Introduction 94
7-2 Tests for Convergence and Divergence 95
7-3 Power Series Revisited 98
7-4 Maclaurin and Taylor Series 99
7-5 Fourier Series and Fourier Transforms 101

8 Scalars and Vectors 110

8-1 Scalars 110
8-2 Vectors and their Addition 111
8-3 Multiplication of Vectors 114
8-4 Applications 117

9 Matrices and Determinants 120

9-1 Introduction 120
9-2 Square Matrices and Determinants 121
9-3 Matrix Algebra 123
9-4 Solution of Systems of Linear Equations 126
9-5 Characteristic Equation of a Matrix 128

10 Operators 135
CH 360

10-1 Introduction 135
10-2 Vector Operators 137
10-3 Eigenvalue Equations Revisited 138
10-4 Hermitian Operators 140
10-5 Rotational Operators 142
10-6 Transformation of ∇^2 to Plane Polar Coordinates 145

11 Numerical Methods and the Use of the Computer 150

11-1 Introduction 150
11-2 Programming a Computer 152
11-3 Programming Examples 163
11-4 Numerical Integration 169
11-5 Roots to Equations 174

12 Mathematical Methods in the Laboratory 179
CH 363-364 Lab

12-1 Introduction 179
12-2 Probability 180
12-3 Experimental Errors 182
12-4 Propagation of Errors 185
12-5 Preparation of Graphs 187
12-6 Tangents and Areas 188

Appendices 192

I Table of Physical Constants 192
II Table of Integrals 192
III Transformation of ∇^2 to Spherical Polar Coordinates 206
IV Stirling's Approximation 208

Answers 210

Index 221

Preface

When I wrote the first edition of *Applied Mathematics for Physical Chemistry* in 1974, I knew of only one other book of this type that was currently available; it was written by Farrington Daniels, published many years earlier and entitled *Mathematics for Physical Chemistry*. It covered basic algebra and calculus and had very little of the advanced mathematics needed to handle so-called modern physical chemistry, which includes quantum chemistry (atomic and molecular structure), spectroscopy, and statistical mechanics. I had found after teaching physical chemistry for a few years that the three semesters of calculus required of our chemistry majors as a prerequisite for taking physical chemistry was just not enough mathematics to do these areas justice. Moreover, I had found that I was taking valuable physical chemistry lecture time to teach my students basic math skills. Not only were students not getting the needed advanced mathematics from the prerequisite mathematics courses, but many of them had difficulty applying the mathematics that they did know to chemical problems. I would like to say that over the past years things have improved, but I find that students are just as unprepared today as they were twenty years ago when the first edition of this book was published. There is still a need for a book such as this to be used along with and to supplement the basic two- or three-semester course in physical chemistry.

Like the first edition, this second edition is in no way intended to replace the calculus courses taken as prerequisite to the physical chemistry course. This text is a *how to do it* review book. While it covers some areas of mathematics in more detail than did the first edition, nevertheless, it concentrates on only those areas of mathematics that are used extensively in physical chemistry, particularly at the undergraduate level. Moreover, it is a mathematics textbook, not a physical chemistry textbook. The primary concern of this book is to help students apply mathematics to chemical problems, and like the first edition, the problems at the ends of the chapters are designed to test the student's mathematical ability, not his or her ability in physical chemistry. I have found that this book is particularly suited to students who have been away from mathematics for several years and are returning to take physical chemistry. Also, students starting graduate school, who may not have had an adequate preparation in advanced calculus, have found this book to be useful.

The first five chapters concentrate on subject matter that normally is covered in prerequisite mathematics courses and should be a review. While it is important to review general and special methods of integration, more emphasis has been placed in this edition on using integral tables for doing standard integration. Like the first edition, and in keeping with its original intent, mathematical rigor was kept at a minimum, giving way to intuition where possible.

The latter half of the book is devoted to those areas of mathematics normally not covered in the prerequisite calculus courses taken for physical chemistry. A number of chapters have been expanded to include material not found in the first edition, but again, for the most part, at the introductory level. For example, the chapter on differential equations expands on the series method of solving differential equations and includes sections on Hermite, Legendre, and Laguerre polynomials; the chapter on infinite series includes a section on Fourier transforms and Fourier series, important today in many areas of spectroscopy; and the chapter on matrices and determinants includes a section on putting matrices in diagonal form, a major type of problem encountered in quantum mechanics.

Finally, a new chapter on numerical methods and computer programming has been included to encourage students to use personal computers to aid them in solving chemical problems. While the programs illustrated in this chapter utilize a more up-to-date form of BASIC, they easily can be modified to other programming languages, such as C. This new chapter concentrates on numerical methods, such as numerical integration, while at the same time showing students how to write simple programs to solve numerical problems that would require long periods of time to solve with the use of only a hand calculator.

No book can be updated without the critical input from professors and students who either used the first edition or would be candidates for using an updated edition. I specifically wish to thank Julie Hutchison, Chapel Hill, North Carolina; Willetta Greene-Johnson, Loyola University, Chicago; Lynmarie A. Posey, Vanderbilt University; Joel P. Ross, St. Michael's College; Sanford A. Safron, Florida State University; and William A. Welsh, University of Missouri for their careful and critical review of the first edition of the text and their many helpful suggestions for the second edition. A special note of thanks is due to my friend and colleague Professor William Porter of the Physics Department here at Southern Connecticut State University for keeping my mathematics honest over these past thirty years.

I thank my editor John Challice, editorial assistant Betsy Williams, total concept coordinator Kimberly Karpovich, copyeditor Trace Wogmon, and the many individuals at Prentice Hall and ETP Harrison whose valuable ideas and comments have helped to improve immensely the quality of this book.

Finally, I wish to thank my son Stephen Barrante, who designed the cover for this edition and who is just beginning his career in graphics design, my wife Marlene, and our family for their patience and encouragement during the preparation of this book.

I welcome comments on the text and ask that these comments or any errors found be sent to me at barrante@pcnet.com.

JAMES R. BARRANTE

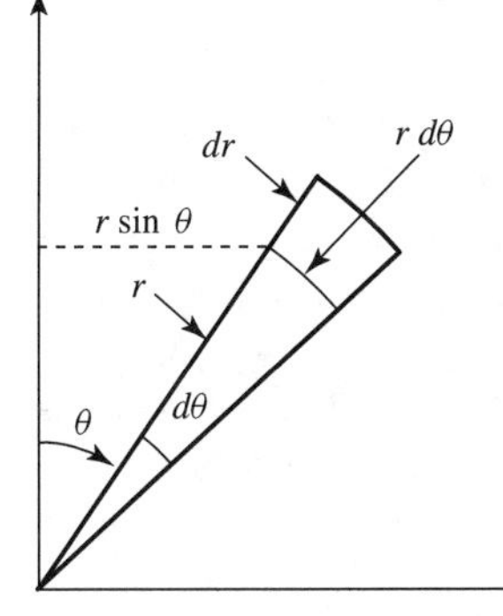

1

Coordinate Systems

1-1 INTRODUCTION

A very useful method for describing the functional dependency of the various properties of a physicochemical system is to assign to each property a point along one of a set of axes, called a *coordinate system*. The choice of coordinate system used to describe the physical world will depend to a great extent on the nature of the properties being described. For example, human beings are very much "at home" in rectangular space. Look around you and note the number of 90° angles you see. Therefore, architects and furniture designers usually employ rectangular coordinates in their work. Atoms and molecules, however, live in "round" space. Generally they operate under potential energy fields that are centrally located and therefore are best described in some form of "round" coordinate system (plane polar or spherical polar coordinates). In fact, many problems in physical chemistry dealing with atoms or molecules cannot be solved in linear coordinates. Thus, as chemists, we need to be familiar with polar as well as linear coordinates. We shall see, however, that things we are accustomed to studying in linear coordinates, such as waves, look very different to us when described in polar coordinates. But, once we have come to terms with polar coordinates, we will find that they are no more difficult to use than linear coordinates. We begin our discussion with a general treatment of linear coordinates.

1-2 CARTESIAN COORDINATES

In the mid-seventeenth century the French mathematician René Descartes proposed a simple method of relating pairs of numbers as points on a rectangular plane surface, today called a *rectangular Cartesian coordinate system*. A typical two-dimensional Cartesian coordinate system consists of two perpendicular axes, called the *coordinate axes*. The vertical or y-axis is called the *ordinate*, while the horizontal or x-axis is called the *abscissa*. The point of intersection between the two axes is called the *origin*. In designating a point on this coordinate system, the abscissa of the point always is given first. Thus, the notation (4, 5) refers to the point whose abscissa is 4 and whose ordinate is 5, as shown in Fig. 1-1.

The application of mathematics to the physical sciences requires taking these abstract collections of numbers and the associated mathematics and relating them to the physical world. Thus, the x's and y's in the above graph could just as easily be pressures or volumes or temperatures describing a gas, or any pair of physical variables that are related to each other.

An example of a Cartesian coordinate system that is used extensively in physical chemistry is illustrated in Fig. 1-2. Here, the ordinate axis represents the variable pressure, while the abscissa represents the variable volume. Since both pressure and volume must be positive numbers, it is customary to omit the negative values from the coordinate system. Any curve drawn on this coordinate system represents the functional dependence of pressure on volume and vice versa. (Functions are described in Chapter 2). For example, the curve shown in the diagram is a representation of Boyle's law, $PV = k$, and describes the inverse proportionality between pressure and volume for an ideal gas. The equation describing this curve on the graph

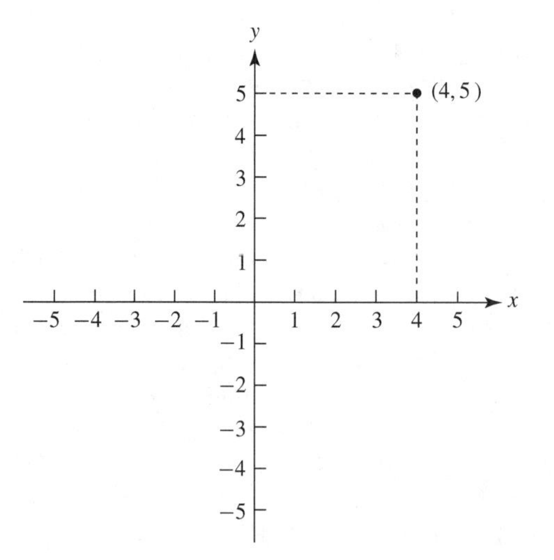

Figure 1-1 Graph of the point (4, 5).

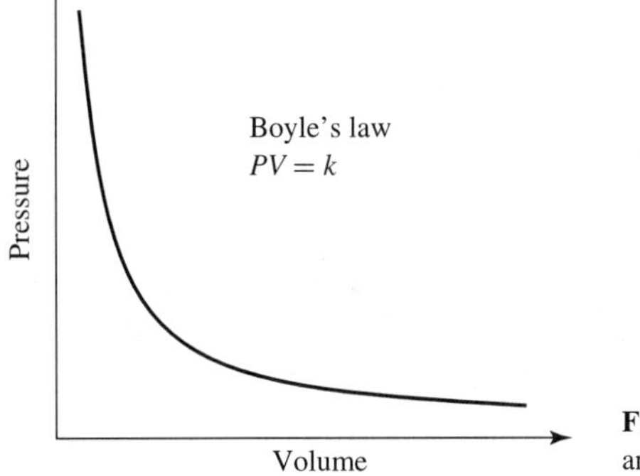

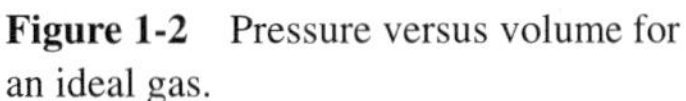
Figure 1-2 Pressure versus volume for an ideal gas.

could just as easily have been written $yx = k$. However, not apparent in either equation ($PV = k$ or $yx = k$) is the fact that this inverse relationship rarely applies to real gases, except at high temperature and low pressure, and then only if the temperature of the gas is held constant. It is a knowledge of facts such as these that continues to distinguish the science of physical chemistry from that of "pure mathematics."

Coordinate systems are not limited to plane surfaces. We can extend the two-dimensional rectangular coordinate system described above into a three-dimensional coordinate system by constructing a third axis perpendicular to the x-y plane and passing through the origin. In fact, we are not limited to three-dimensional coordinate systems and can extend the process to as many coordinates as we wish or need—n-dimensional coordinate systems—although, as we might expect, the graphical representation becomes difficult beyond three dimensions, since we are creatures of three-dimensional space. For example, in wave mechanics, an area of physics describing the mechanical behavior of waves, we usually describe the amplitude of a transverse wave by using an "extra" coordinate. Therefore, the amplitude of a one-dimensional transverse wave—say, along the x-axis—is represented along the y-axis, illustrated for a simple sine wave in Fig. 1-3. (Keep in mind that the wave itself does not move in the y-direction. The wave is a disturbance, or series of disturbances, traveling, in this case, only down the x-axis. We should not confuse the motion of the wave with the motion of the medium producing the wave.) A two-dimensional transverse wave in the xy-plane, such as a water wave moving across the surface of a lake, requires a third coordinate, call it the z-axis, to describe the amplitude of the wave—still easy to represent graphically, and certainly within the realm of our experience. Many problems in quantum wave mechanics, however, require us to describe three-dimensional waves. In order to describe graphically the amplitude of a three-dimensional transverse wave, we need a fourth coordinate. This presents a problem to creatures of three-dimensional space, and a number of ingenious ways have been devised to get around the problem. One way is to use the density of points on a three-dimensional graph to represent the amplitude of the wave.

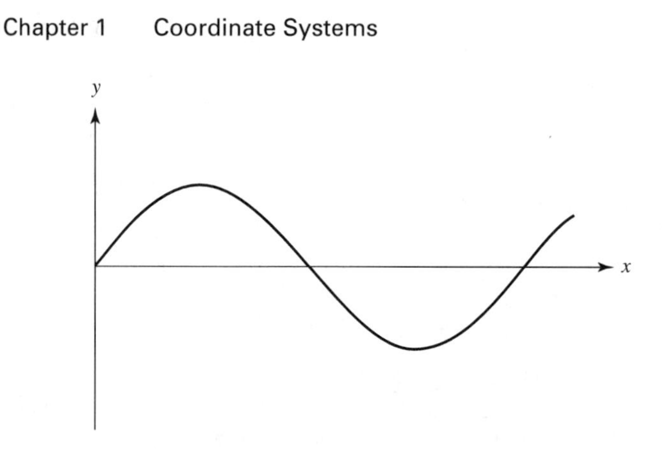

Figure 1-3 One-dimensional sine wave traveling down x-axis.

In any of the coordinate systems described above, it is useful to define a very small or differential volume element

$$d\tau = dq_1\, dq_2\, dq_3\, dq_4 \ldots dq_n \tag{1-1}$$

where dq_i is an infinitesimally small length along the ith axis. In the three-dimensional Cartesian coordinate system shown in Fig. 1-4, the volume element is simply

$$d\tau = dx\, dy\, dz \tag{1-2}$$

1-3 PLANE POLAR COORDINATES

Many problems in the physical sciences cannot be solved in rectangular space. For this reason, we find it necessary to redefine the Cartesian axes in terms of what nor-

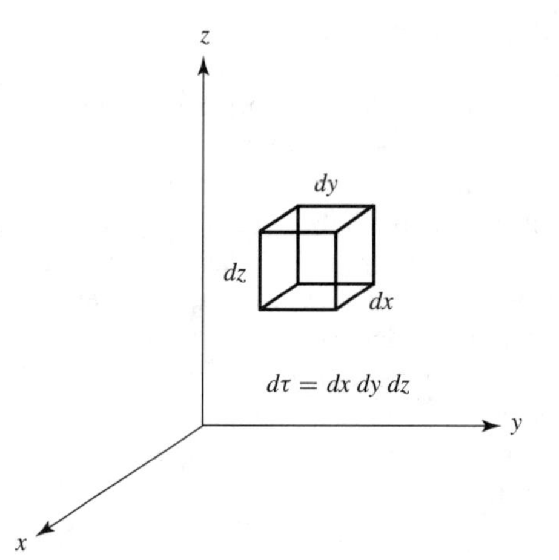

Figure 1-4 Differential volume element for a Cartesian coordinate system.

Applied Mathematics for Physical Chemistry

SECOND EDITION

JAMES R. BARRANTE

Department of Chemistry
Southern Connecticut State University

PRENTICE HALL
Upper Saddle River, New Jersey 07458

Library of Congress Cataloging-in-Publication Data

Barrante, James R.
Applied mathematics for physical chemistry / James R. Barrante. — 2nd ed.
p. cm.
Includes index.
ISBN 0-13-741737-3
1. Chemistry, Physical and theoretical—Mathematics. I. Title.
QD455.3.M3B37 1998
510′.24′541—dc21 97-17387
CIP

Acquisitions editor: John Challice
Total concept coordinator: Kimberly P. Karpovich
Manufacturing manager: Trudy Pisciotti
Production: ETP Harrison
Copy editor: Trace Wogmon
Cover designer: Stephen Barrante

Simon & Schuster/A Viacom Company
Upper Saddle River, New Jersey 07458

Printed in the United States of America

10 9 8 7 6 5 4 3 2 1

ISBN 0-13-741737-3

Prentice-Hall International (UK) Limited, *London*
Prentice-Hall of Australia Pty. Limited, *Sydney*
Prentice-Hall Canada Inc., *Toronto*
Prentice-Hall Hispanoamericana, S.A., *Mexico*
Prentice-Hall of India Private Limited, *New Delhi*
Prentice-Hall of Japan, Inc., *Tokyo*
Simon & Schuster Asia Pte. Ltd., *Singapore*
Editora Prentice-Hall do Brasil, Ltda., *Rio de Janeiro*

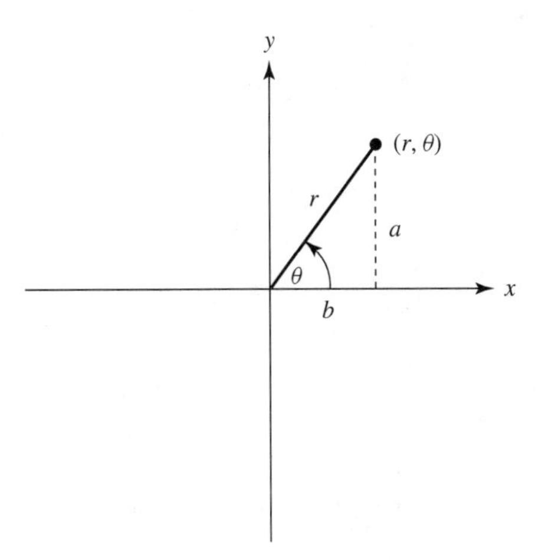

Figure 1-5 Plane polar coordinates.

mally are referred to as "round" or "curvilinear" coordinates. Consider the diagram in Fig. 1-5. It is possible to associate every point on this two-dimensional coordinate system with the geometric properties of a right triangle. Note that the magnitude of x is the same as the length of side b of the triangle shown, and that the magnitude of y is the same as the length of side a. Since

$$\sin\theta = \frac{a}{r} \quad \text{and} \quad \cos\theta = \frac{b}{r} \tag{1-3}$$

we can write

$$x = b = r\cos\theta \quad \text{and} \quad y = a = r\sin\theta \tag{1-4}$$

Therefore, every point (x, y) can be specified by assigning to it a value for r and a value for θ. This type of graphical representation is called a *plane polar coordinate system.* In this coordinate system, points are designated by the notation (r, θ). The Equations (1-4) are known as transformation equations; they transform the coordinates of a point from polar coordinates to Cartesian (linear) coordinates. The reverse transformation equations can be found by simple trigonometry.

$$\frac{y}{x} = \frac{r\sin\theta}{r\cos\theta} = \tan\theta \quad \text{or} \quad \theta = \tan^{-1}\left(\frac{y}{x}\right) \tag{1-5}$$

and

$$r = (x^2 + y^2)^{1/2} \tag{1-6}$$

1-4 SPHERICAL POLAR COORDINATES

One extension of the plane polar coordinate system to three dimensions leads to the *spherical polar coordinate system*, shown in Fig. 1-6. A point in this system is

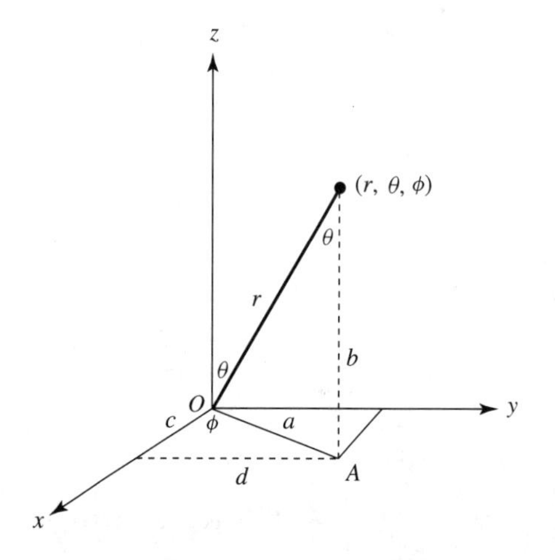

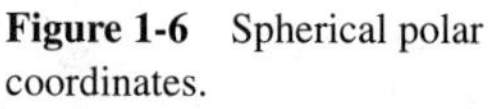
Figure 1-6 Spherical polar coordinates.

represented by three numbers: r, the distance of the point from the origin; θ, the angle that the line r makes with the z-axis; and ϕ, the angle that the line $\overline{OA}$ makes with the x-axis. Since

$$\cos\phi = \frac{c}{a}, \quad \sin\phi = \frac{d}{a}, \quad \text{and} \quad \sin\theta = \frac{a}{r} \tag{1-7}$$

and since the length of c is numerically equal to x, d to y, and b to z, we can write

$$\begin{aligned} x &= a\cos\phi = r\sin\theta\cos\phi \\ y &= a\sin\phi = r\sin\theta\sin\phi \\ z &= r\cos\theta \end{aligned} \tag{1-8}$$

The reverse transformation equations are found as follows:

$$\frac{y}{x} = \frac{r\sin\theta\sin\phi}{r\sin\theta\cos\phi} = \tan\phi \quad \text{or} \quad \phi = \tan^{-1}\left(\frac{y}{x}\right) \tag{1-9}$$

$$r = (x^2 + y^2 + z^2)^{1/2} \tag{1-10}$$

$$\cos\theta = \frac{z}{r} \quad \text{or} \quad \theta = \cos^{-1}\frac{z}{(x^2 + y^2 + z^2)^{1/2}} \tag{1-11}$$

The differential volume element in spherical polar coordinates is not as easy to determine as it is in linear coordinates. Recalling, however, that the length of a circular arc intercepted by an angle θ is $L = r\theta$, where r is the radius of the circle, we can see from Fig. 1-7 that the volume element is

$$d\tau = r^2 \sin\theta \, dr \, d\theta \, d\phi \tag{1-12}$$

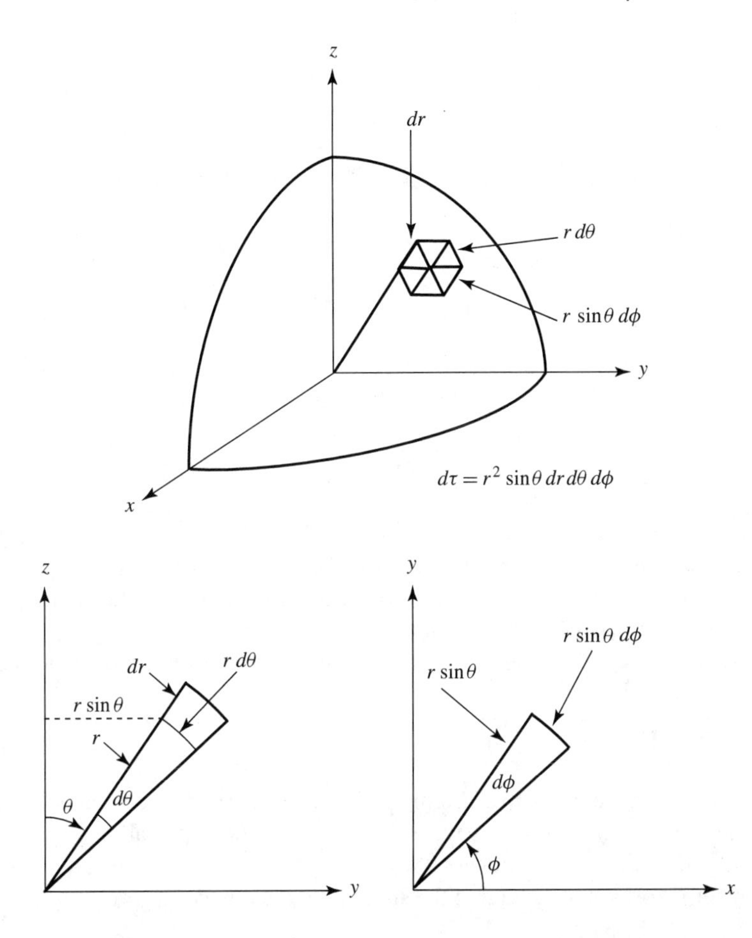

Figure 1-7 Differential volume element in spherical polar coordinates.

1-5 THE COMPLEX PLANE

A complex number is a number composed of a real part x and an imaginary part iy, where $i = \sqrt{-1}$, and normally is represented by the equation $z = x + iy$. A real number, then, is one in which $y = 0$, while a pure imaginary number is one in which $x = 0$. Thus, in a sense, all numbers can be thought of as complex numbers.

It is possible to represent complex numbers by means of a coordinate system. The real part of the complex number is designated along the x-axis, while the pure imaginary part is designated along the y-axis, as shown in Fig. 1-8. Since $x = r\cos\theta$ and $y = r\sin\theta$, any complex number can be written as

$$z = x + iy = r\cos\theta + ri\sin\theta = r(\cos\theta + i\sin\theta) \tag{1-13}$$

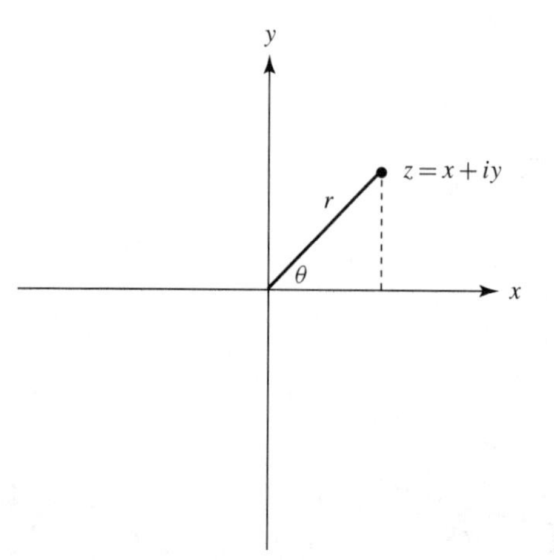

Figure 1-8 The complex plane.

Moreover, since every point in the plane formed by the x- and y-axes represents a complex number, the plane is called the *complex plane*. In an n-dimensional coordinate system, one plane may be the complex plane.

The value of r in Equation (1-13), called the *modulus* or *absolute value*, can be found by the equation

$$r = (x^2 + y^2)^{1/2} = |z| \tag{1-14}$$

The angle θ, called the *phase angle*, simply describes the rotation of z in the complex plane. The square of the absolute value can be shown to be identical to the product of $z = x + iy$ and its complex conjugate $z^* = x - iy$. The complex conjugate of a complex number is formed by changing the sign of the imaginary part.

$$z^*z = (x - iy)(x + iy) = x^2 + y^2 = |z|^2 \tag{1-15}$$

To find another useful relationship between sin θ and cos θ in the complex plane, let us expand each function in terms of a Maclaurin series. (Series expansions are covered in detail in Chapter 7.)

$$\sin\theta = \theta - \frac{\theta^3}{3!} + \frac{\theta^5}{5!} - \frac{\theta^7}{7!} + - \cdots \tag{1-16}$$

$$\cos\theta = 1 - \frac{\theta^2}{2!} + \frac{\theta^4}{4!} - \frac{\theta^6}{6!} + - \cdots \tag{1-17}$$

where $n!$ (n-factorial) is $n! = 1(2)(3)(4) \ldots (n - 1)(n)$. Hence,

$$\cos\theta + i\sin\theta = 1 + i\theta - \frac{\theta^2}{2!} - \frac{i\theta^3}{3!} \cdots \tag{1-18}$$

But this is identical to the series expansion for $e^{i\theta}$

$$e^{i\theta} = 1 + i\theta - \frac{\theta^2}{2!} - \frac{i\theta^3}{3!} \cdots \tag{1-19}$$

Hence, we can write

$$e^{i\theta} = \cos\theta + i\sin\theta \tag{1-20}$$

and

$$z = r\,e^{i\theta} \tag{1-21}$$

It can be shown by the same method used above that

$$e^{-i\theta} = \cos\theta - i\sin\theta \tag{1-22}$$

and

$$z^* = r\,e^{-i\theta} \tag{1-23}$$

SUGGESTED READING

1. BRADLEY, GERALD L., and SMITH, KARL J., *Calculus*, Prentice-Hall, Inc., Upper Saddle River, NJ, 1995.
2. SULLIVAN, MICHAEL, *College Algebra,* 4th ed., Prentice-Hall, Inc., Upper Saddle River, NJ, 1996.
3. VARBERG, DALE, and PURCELL, EDWIN J., *Calculus*, 7th ed., Prentice-Hall, Inc., Upper Saddle River, NJ, 1997.
4. WASHINGTON, ALLYN J., *Basic Technical Mathematics*, 6th ed., Addison-Wesley Publishing Co., Boston, 1995.

PROBLEMS

1. What is the sign of the abscissa and ordinate of points in each of the quadrants of a two-dimensional coordinate system? Relate these to the sign of $\sin\theta$, $\cos\theta$, and $\tan\theta$ in each quadrant.

2. Determine the values of r and θ for the following points:

(a) $(2, 2)$ **(d)** $(4, -1)$ **(g)** $(-2, 0)$
(b) $(1, \sqrt{2})$ **(e)** $(-3, -2)$ **(h)** $(0, -5)$
(c) $(1, 5)$ **(f)** $(2, 0)$ **(i)** $(12, -6)$

3. Determine the values of x and y for the following points:

(a) $r = 1.11, \theta = 54°22'$ **(e)** $r = 6.00, \theta = 145°$
(b) $r = 1.00, \theta = 0$ **(f)** $r = 2.50, \theta = 270°$
(c) $r = 3.16, \theta = 225°$ **(g)** $r = 3.00, \theta = 35°$
(d) $r = \sqrt{3}, \theta = 90°$ **(h)** $r = 5.00, \theta = 71°34'$

4. Determine the values of r, θ, and ϕ for the following points:

(a) $(1, 1, 1)$ **(c)** $(2, 0, -1)$ **(e)** $(-3, -6, -12)$
(b) $(3, 2, 1)$ **(d)** $(-1, 0, 4)$ **(f)** $(0, 0, -4)$

5. The *cylindrical coordinate system* can be constructed by extending a z-axis from the origin of a plane polar coordinate system perpendicular to the x-y plane. A point in this system is designated by the coordinates (r, θ, z). What is the differential volume element in this coordinate system?

6. Determine the modulus and phase angle for the following complex numbers:

(a) 3 **(c)** $2 + 2i$ **(e)** $-4 - 4i$

(b) $6i$ **(d)** $1 - 3i$ **(f)** $-4 + 5i$

7. Show that $e^{-i\theta} = \cos\theta - i\sin\theta$.

8. Show that $\cos\theta = \frac{1}{2}(e^{i\theta} + e^{-i\theta})$ and $\sin\theta = \frac{1}{2i}(e^{i\theta} - e^{-i\theta})$.

9. Find the values of m that satisfy the equation $e^{2\pi im} = 1$. (*Hint:* Express the exponential in in terms of sines and cosines.)

10. Show that $A\,e^{ikx} + B\,e^{-ikx}$, where A and B are arbitrary constants, is equivalent to the sum $A'\sin kx + B'\cos kx$, where A' and B' are arbitrary constants.

11. A *space-time* diagram is a two-dimensional coordinate system in which position is plotted on one axis (usually the y-axis) and time is plotted on the other (usually on the x-axis). A line on this coordinate system, called a *world line*, represents motion of a particle through space and time. Construct world lines on a space-time diagram showing the following:

(a) a particle at rest relative to the observer

(b) a particle moving slowly relative to the observer

(c) a particle moving very fast relative to the observer

(d) Would a vertical line be possible on this coordinate system?

2

Functions and Graphs

2-1 FUNCTIONS

Physical chemistry, like all the physical sciences, is concerned with the dependence of one or more variables of a system upon other variables of the system. For example, suppose we wished to know how the volume of a gas varies with temperature. With a little experimentation in the laboratory, we would find that the volume of a gas varies with temperature in a very specific way. Careful measurements would show that the volume of a gas V_t at any temperature t on the Celsius scale obeys the specific law

$$V_t = V_0(1 + \alpha t) \tag{2-1}$$

where V_0 is the volume of the gas at 0°C and α is a constant known as the *coefficient of expansion* of the gas. This equation predicts that there is a one-to-one correspondence between the volume of a gas and its temperature. That is, for every value of t substituted into Equation (2-1), there is a corresponding value for V_t.

Let us define a collection of temperatures as a mathematical set $T = \{t_1, t_2, t_3, t_4, \ldots\}$ and the corresponding volumes as another set $V = \{V_1, V_2, V_3, V_4, \ldots\}$. A mathematical set is defined as a collection of numbers, each member of the set called an *element*, so we see that our collection of temperatures and volumes satisfies this definition. If there is associated with each element of set T at least one element in the other set V, then this association is said to constitute a function from T to V, written $f\colon t \rightarrow V_t$. That is, the function takes every element in set T into the corresponding element in set V. We see that Equation (2-1) satisfies this condition. Since V_t is a function of t—that is, the value of V depends on the value of t—the above expression can be written $f: t \rightarrow f(t)$, where $V_t = f(t)$. Remember that $f(t)$, read "f of t," does not mean f is multiplied by t, but that $f(t)$ is the value of $V_0(1 + \alpha t)$ at t. Hence, we can write

$$f(t) = V_0(1 + \alpha t) \tag{2-2}$$

A function, then, is defined as a correspondence between elements of two mathematical sets.

In the above example, V_t was considered to be a function of only a single variable t. Such an equation, $V_t = f(t)$, can be represented by a series of points on a two-dimensional Cartesian coordinate system. Physicochemical systems, however, usually depend on more than one variable. Thus, it is necessary to extend the definition of function given above to include functions of more than one variable. For example, we find experimentally that the volume of a gas will vary with temperature according to Equation (2-2) only if the pressure of the gas is held constant. Thus, the volume of a gas is not only a function of temperature, but also is a function of pressure. Careful measurements in the laboratory will show that for most gases at or around room temperature and one atmosphere pressure the law relating the volume of a gas simultaneously to the temperature and the pressure of the gas is the well-known ideal gas law

$$V = \frac{RT}{P} = f(T, P) \tag{2-3}$$

where R is a constant. Equation (2-3) implies that there is a one-to-one correspondence between three sets of numbers: a set of volumes, $V = \{V_1, V_2, V_3, V_4, \ldots\}$; a set of temperatures on the absolute temperature scale, $T = \{T_1, T_2, T_3, T_4, \ldots\}$; and a set of pressures, $P = \{P_1, P_2, P_3, P_4, \ldots\}$. These three sets can be represented graphically on a three-dimensional coordinate system by plotting V along one axis, T along a second axis, and P along the third axis. Such graphs of P, V, and T commonly are called *phase diagrams*.

2-2 GRAPHICAL REPRESENTATION OF FUNCTIONS

As we saw above, one of the most convenient ways to represent the functional dependence of the variables of the system is by the use of coordinate systems. This is because each set of numbers is easily represented by a coordinate axis, and the graphs that result give an immediate visual representation of the behavior. In this section we shall explore several types of graphical representation of functions. We begin with functions that describe a linear dependence between the variables.

Equations of the First Degree. Equations that define functions showing a linear dependence between variables are known as *equations of the first degree*, or *first-degree equations*. These functions describe a dependence commonly called the *direct proportion*, and the equations are called first-degree equations because all the variables in these equations are raised only to the first power. First-degree equations have the general form

$$f(x) = mx + b \tag{2-4}$$

where m and b are constants. As an example of a first-degree equation, consider the equation $°F = \frac{9}{5}°C + 32$, illustrated in Fig. 2-1. This familiar equation describes the temperature of a system on the Fahrenheit scale and its relationship to the temperature on the Celsius temperature scale. Note that, indeed, the equation is that of a straight line. From the graph we can determine the significance of the constants m and b in Equation (2-4). Let us consider the latter, $b = 32$. We can see that the line crosses the $°F$-axis at the point where the value of $°C$ is zero. If we substitute zero for $°C$ into the equation $°F = \frac{9}{5}°C + 32$, we obtain $°F = 32$. But this is just the value of b in the equation. Hence, $(0, b)$ are the coordinates of the point where the line crosses the $°F$- or y-axis, and thus b is called the *y-intercept*. It also may be of interest to consider the value of $°C$ for which $°F = f(°C) = 0$. This point represents the point where the line crosses the $°C$- or x-axis, and is known as the *zero of the function*. Rearranging Equation (2-4) gives

$$0 = mx + b \quad \text{or} \quad x = -\frac{b}{m} \tag{2-5}$$

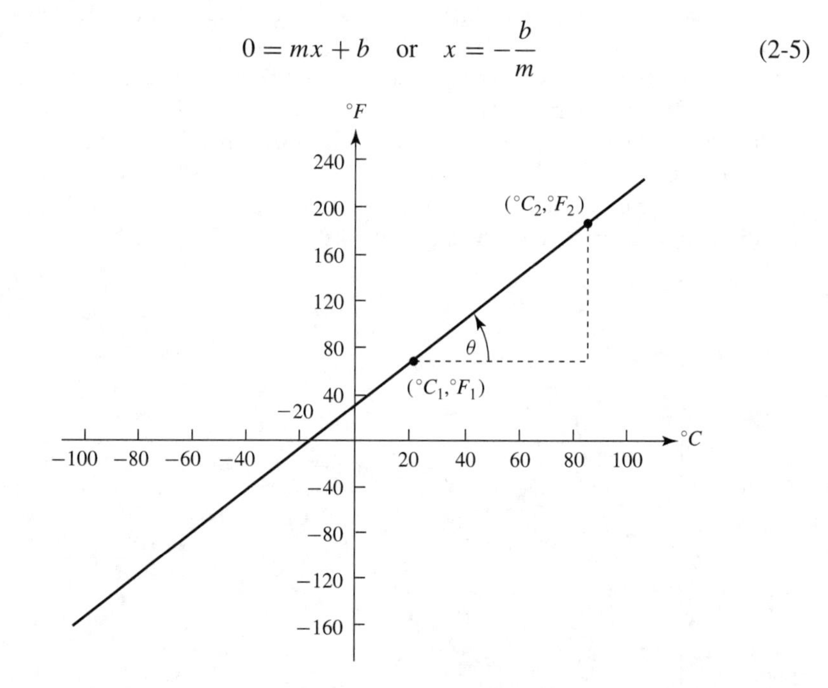

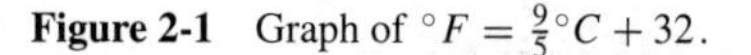
Figure 2-1 Graph of $°F = \frac{9}{5}°C + 32$.

Let us consider, next, any two points on the line—say $(^\circ C_1, ^\circ F_1)$ and $(^\circ C_2, ^\circ F_2)$—where

$$^\circ F_1 = \frac{9}{5}{}^\circ C_1 + 32 \quad \text{and} \quad ^\circ F_2 = \frac{9}{5}{}^\circ C_2 + 32 \tag{2-6}$$

We can define the slope of the line as the ratio of the change in y, $\Delta y = y_2 - y_1$, to the change in x, $\Delta x = x_2 - x_1$. But this is just the tangent of the angle θ (shown in Fig. 2-1). Hence, we can write

$$\text{slope} = \frac{\Delta y}{\Delta x} = \frac{y_2 - y_1}{x_2 - x_1} = \tan\theta \tag{2-7}$$

Substituting Equation (2-6) into Equation (2-7) gives

$$\text{slope} = \frac{\Delta^\circ F}{\Delta^\circ C} = \frac{\frac{9}{5}{}^\circ C_2 + 32 - \frac{9}{5}{}^\circ C_1 - 32}{^\circ C_2 - {}^\circ C_1} = \frac{9}{5} \tag{2-8}$$

This, however, is just the value of m in the equation. A little experience with algebra will show that the constant m is always the slope of the line found by measuring the tangent of the angle that the line makes with the x-axis.

In many cases, equations that do not exhibit a linear relationship between the variables can be made to do so by simply choosing the correct set of coordinate axes on which to plot them. It is useful to do so, because the physical constants in an equation are most easily determined graphically, if the equation is linear. For example, the graph of Boyle's law, P versus V, shown in Fig. 1-2, is hardly linear. If, however, P versus $1/V$ rather than P versus V is plotted, the straight line shown in Fig. 2-2 results. The constant k, easily found to be the slope of the line in this case, would have been very difficult to determine graphically from the graph in Fig. 1-2.

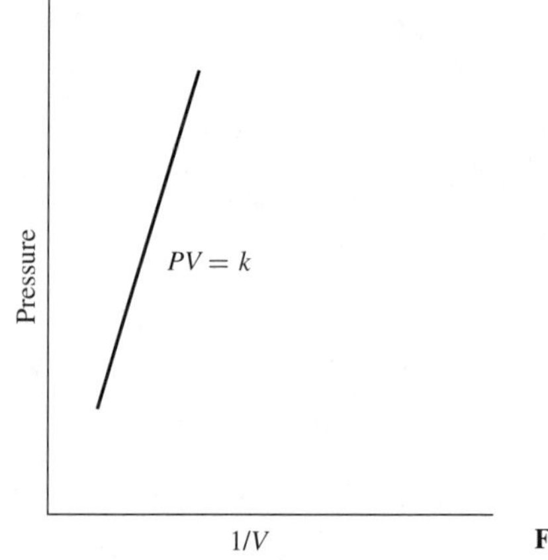

Figure 2-2 Graph of $PV = k$.

Equations of the Second Degree. Equations of the second degree are those having the general form

$$f(x) = ax^2 + bx + c \tag{2-9}$$

where a, b, and c are constants and $a \neq 0$. Thus, a second-degree equation is one in which the highest power to which the variable x is raised is 2. Equation (2-9) is given the specific name *quadratic equation*, and the function that it defines is called a *quadratic function*. An example of a typical quadratic equation, $y = x^2 - 2x - 2$, is illustrated in Fig. 2-3. The family of curves that Equation (2-9) describes are called *parabolas*. We can see by experience with this type of curve that when the constant a is positive, the curve opens upward, and when the constant a is negative, the curve opens downward. Notice that the slope of the curve at the point (x, y), which can be defined by a line drawn tangent to the curve at this point, is not constant, but changes as a function of x, as shown in Fig. 2-4. In practice, this type of slope is very difficult to determine graphically, since tangent lines to the curve must be constructed. However, we shall see in Chapter 4 that differential calculus can be used to determine the slope of the curve at any point, provided the equation describing the curve is known.

In the previous section we defined the zero of the function as the point where a line crosses the x-axis. In the case of parabolas, there must be two zeros of the function (although in some cases the two zeros might be identical). In order to determine these zeros, we must solve the equation

$$y = ax^2 + bx + c = 0 \tag{2-10}$$

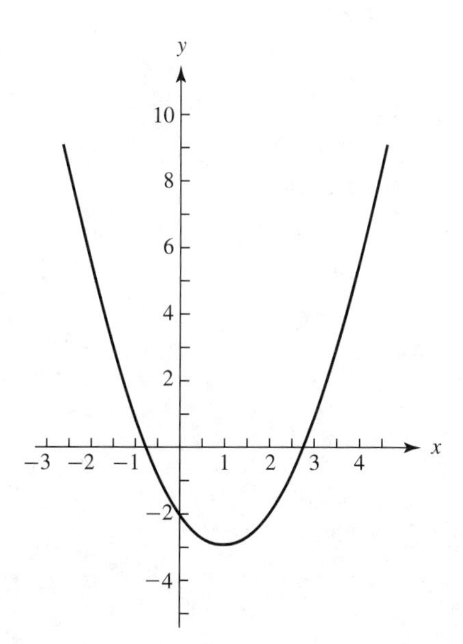

Figure 2-3 Graph of $y = x^2 - 2x - 2$.

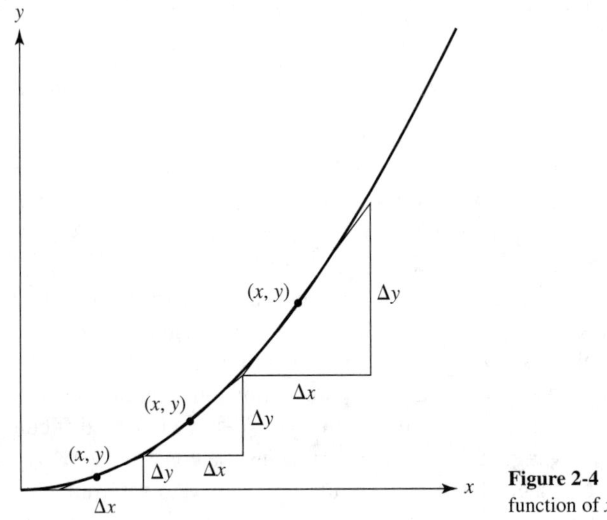

Figure 2-4 Variation of slope as a function of x.

for the values of x. Dividing both sides of the equation by a and rearranging the equation gives

$$x^2 + \frac{b}{a}x = -\frac{c}{a} \tag{2-11}$$

Next, adding $b^2/4a^2$ to both sides of the equation to complete the square gives

$$x^2 + \frac{b}{a}x + \frac{b^2}{4a^2} = \frac{b^2}{4a^2} - \frac{c}{a} \tag{2-12}$$

or

$$\left(x + \frac{b}{2a}\right)^2 = \frac{b^2 - 4ac}{4a^2}$$

Taking the square root of both sides of the equation gives

$$\left(x + \frac{b}{2a}\right) = \frac{\pm\sqrt{b^2 - 4ac}}{2a} \tag{2-13}$$

or

$$x = \frac{-b \pm \sqrt{b^2 - 4ac}}{2a} \tag{2-14}$$

which is the well-known *quadratic formula*.

Sometimes the zeros of the equation, called the *roots*, can be determined by the factoring method. For example, consider the equation

$$x^2 - 3x + 2 = 0 \tag{2-15}$$

which can be factored into the terms

$$(x - 1)(x - 2) = 0 \tag{2-16}$$

The roots of the equation now can be found by solving the equations

$$(x - 1) = 0 \quad \text{and} \quad (x - 2) = 0 \tag{2-17}$$

which gives $x = 1$ and $x = 2$. Substituting $a = 1$, $b = -3$, and $c = 2$ into the quadratic formula, Equation (2-14), yields the same results.

In cases where the equation defining a particular physical situation is a second-degree equation (or even one of higher order), there arises a problem that is not present when one simply considers the pure mathematics, as we have done above. Since quadratic equations necessarily have two roots, we must decide, in cases where both roots are not the same, which root correctly represents the physical situation, even though both are mathematically correct. For example, consider the equilibrium equation

$$A + B \rightleftharpoons C + D$$

Assume that initially the concentrations of A, B, C, and D are each 1 molar. Suppose we wish to determine the equilibrium concentrations of A, B, C, and D given that the equilibrium constant in terms of molar concentrations, K_c, equals 50. If we assume that at equilibrium the concentration of C is $(1 + x)$ molar, then the equilibrium concentrations of A, B, and D must be $(1 - x)$, $(1 - x)$, and $(1 + x)$ molar, respectively. Substituting these values into the equilibrium constant equation

$$K_c = \frac{(C)(D)}{(A)(B)}$$

we have

$$50 = \frac{(1 + x)(1 + x)}{(1 - x)(1 - x)} \tag{2-18}$$

Rearranging Equation (2-18) gives the quadratic equation

$$49x^2 - 102x + 49 = 0 \tag{2-19}$$

Substituting the values $a = 49$, $b = -102$, and $c = 49$ into Equation (2-14) yields the two solutions $x = 1.3$ and $x = 0.75$.

We now must decide which value of x is physically correct. If we choose $x = 1.3$, the equilibrium concentrations of A and B will be negative numbers, which physically does not make sense. Thus, the physically correct value for x must be 0.75, giving for the equilibrium concentrations of A, B, C, and D: $0.25M$, $0.25M$, $1.75M$, and $1.75M$, respectively. We see, then, that although both roots were mathematically correct, only one root made sense physically.

Exponential and Logarithmic Functions. Exponential functions are functions whose defining equation is written in the general form

$$f(x) = a^x \tag{2-20}$$

where $a > 0$. An important exponential function that is used extensively in physical chemistry, and indeed in the physical sciences as a whole, is the function

$$f(x) = e^x \tag{2-21}$$

where the constant e is a nonterminating, nonrepeating decimal having the value, to five significant figures,

$$\begin{aligned} e &= \lim_{x \to 0} (1 + x)^{1/x} = 1 + \frac{1}{1!} + \frac{1}{2!} + \frac{1}{3!} + \cdots \\ &= 2.7183 \end{aligned}$$

This function is illustrated in Fig. 2-5. Note that all exponentials have the point (0,1) in common, since $a^0 = 1$ for any a. Also note that there are no zeros to the function, since the function approaches zero as x approaches $-\infty$. The expression $\lim_{x \to 0}$ means that $(1 + x)^{1/x}$ approaches a value of 2.7183 as x approaches 0, and is read "in the limit that x approaches zero." The physical significance of the constant e will be discussed in Chapter 3.

There is a direct relationship between exponential functions and logarithmic functions. The power or exponent to which the constant a is raised in the equation $y = a^x$ is called the logarithm of y to the base a and is written

$$\log_a y = x \tag{2-22}$$

The logarithmic function $\log_2 y = x$ is illustrated in Fig. 2-6. Note, as in the case of exponential functions, that the point (0,1) is common to all logarithmic functions, since $\log_a 1 = 0$ for any a. Logarithms have many useful properties and are an important tool in the study of physical chemistry. For this reason the general properties of logarithms are reviewed in Chapter 3.

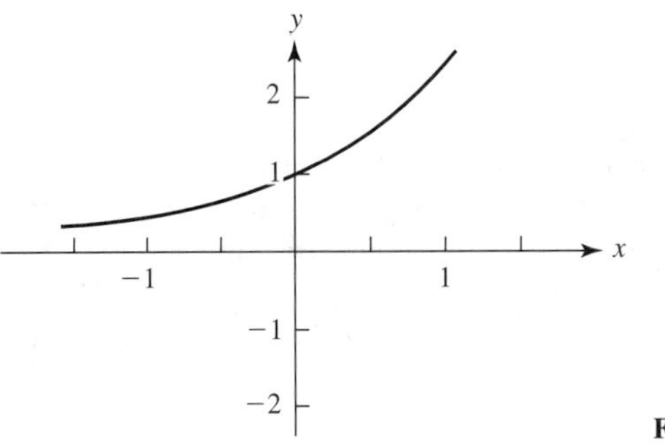

Figure 2-5 Graph of $y = e^x$.

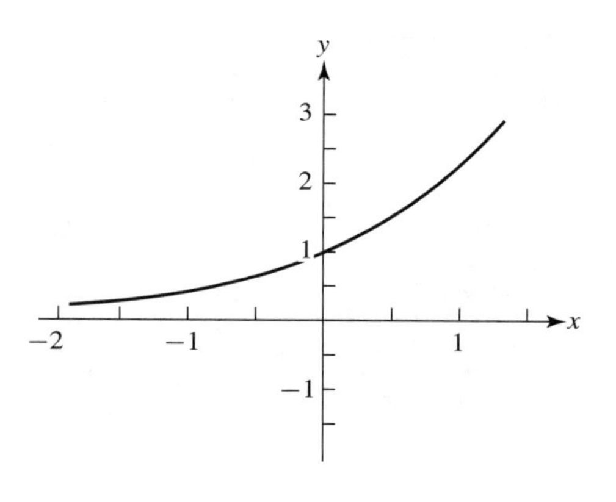

Figure 2-6 Graph of $\log_2 y = x$.

Circular Functions. A circle is defined as the locus of all points in a plane that are at a constant distance from a fixed point. Circles are described by the equation

$$(x - a)^2 + (y - b)^2 = r^2 \tag{2-23}$$

where a and b are the coordinates of the center of the circle (the fixed point) and r is the radius. A unit circle is one with its center at the origin and a radius equal to unity.

$$x^2 + y^2 = 1 \tag{2-24}$$

Consider, now, the triangle inscribed in the unit circle shown in Fig. 2-7. Let us define three functions: sine (abbreviated sin), which takes the angle θ into the y-coordinate of a point (x, y), cosine (abbreviated cos), which takes the angle θ into the x-coordinate of the point (x, y), and tangent (abbreviated tan), which is the ratio of y

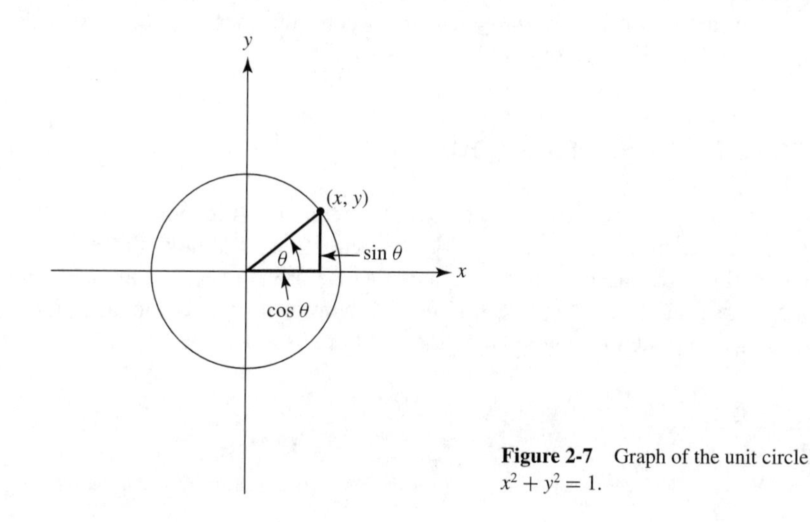

Figure 2-7 Graph of the unit circle $x^2 + y^2 = 1$.

TABLE 2-1 DEPENDENCE OF r ON θ FOR THE FUNCTION $r = A\cos\theta$

θ (degrees)	r	θ (degrees)	r	θ (degrees)	r
0	1.000A	135	−0.707A	270	0
30	0.866A	150	−0.866A	300	0.500A
45	0.707A	180	−1.000A	315	0.707A
60	0.500A	210	−0.866A	330	0.866A
90	0	225	−0.707A	360	1.000A
120	−0.500A	240	−0.500A		

to x. Thus,

$$\begin{aligned} \sin\theta &= y \\ \cos\theta &= x \\ \tan\theta &= \frac{y}{x} = \frac{\sin\theta}{\cos\theta} \end{aligned} \tag{2-25}$$

These functions are called *circular* or *trigonometric* functions. Note that Equations (2-25) are just the transformation Equations (1-4) with $r = 1$. It is interesting to compare the graphs of functions, such as $\sin\theta$ and $\cos\theta$, in linear coordinates (coordinates in which θ is plotted along one axis) to those in plane polar coordinates. Consider, for example, the equation $r = A\cos\theta$, where A is a constant. Such an equation can be used to describe the wave properties of p-type atomic orbitals in two dimensions. The functional dependence of r upon θ can be seen in Table 2-1.

When r versus θ is plotted in linear coordinates [shown in Fig. 2-8(a)], the typical cosine curve results. On the other hand, when r versus θ is plotted in polar coordinates [shown in Fig. 2-8(b)], the more familiar shape of the p-orbital can be seen.[1] It is important to note that both graphs are equivalent, the shapes of the curves depending merely on the choice of coordinate system.

2-3 ROOTS TO POLYNOMIAL EQUATIONS

We saw in the previous sections that the zeros of the function (the roots) can be found easily if the equations are first- or second-degree equations. But how do we find the roots to equations that are not linear or quadratic? Before the age of computers this was not a simple task. One standard way to find the roots of a polynomial equation without using a computer is to graph the function. For example, consider the equation

$$y = x^4 + x^3 - 3x^2 - x + 1$$

[1]In polar coordinates, negative values of r have no meaning, so we are actually plotting $|r| = A\cos\theta$.

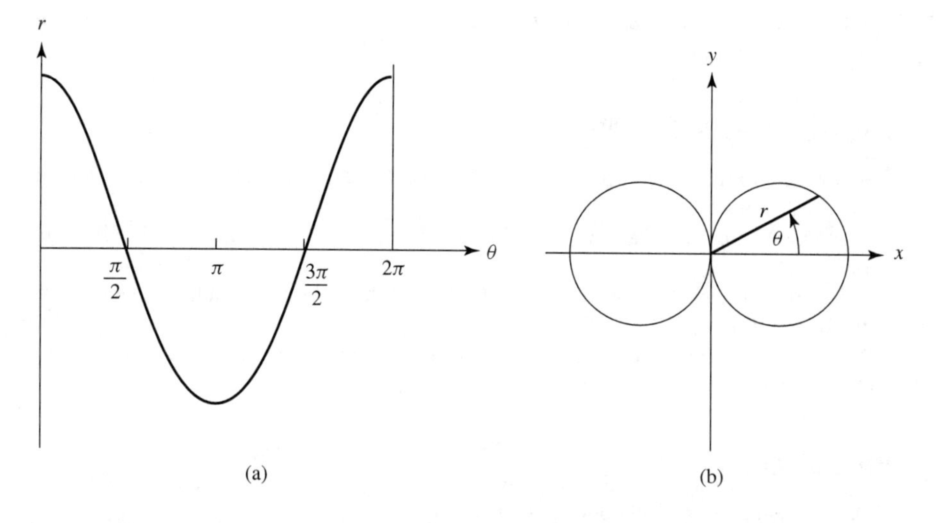

Figure 2.8 Graphs of $r = A \cos \theta$ plotted in (a) linear coordinates and (b) polar coordinates.

If we plot this function from $x = -3$ to $x = +3$, we obtain the graph shown in Fig. 2-9. The roots to the equation are the values of x for which $y = 0$, or the points on graph where the graph crosses the x-axis. Careful examination of the graph will show that the roots are $x = -2.09$, $x = -0.74$, $x = +0.47$, and $x = +1.36$. In

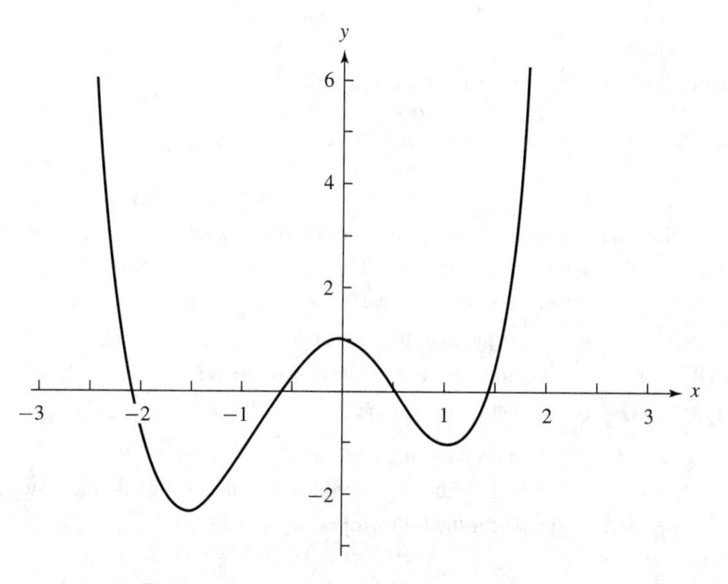

Fig. 2-9 Graph of $y = x^4 + x^3 - 3x^2 - x + 1$.

Chapter 11 we shall discuss numerical methods of finding roots to polynomial equations using a computer.

SUGGESTED READINGS

1. BRADLEY, GERALD L., and SMITH, KARL J., *Calculus*, Prentice-Hall, Inc., Upper Saddle River, NJ, 1995.
2. SULLIVAN, MICHAEL, *College Algebra,* 4th ed., Prentice-Hall, Inc., Upper Saddle River, NJ, 1996.
3. VARBERG, DALE, and PURCELL, EDWIN J., *Calculus*, 7th ed., Prentice-Hall, Inc., Upper Saddle River, NJ, 1997.
4. WASHINGTON, ALLYN J., *Basic Technical Mathematics*, 6th ed., Addison-Wesley Publishing Co., Boston, 1995.

PROBLEMS

1. Determine the zeros of the following functions:
(a) $y = 5x - 5$
(b) $3(y - 1) = -6x$
(c) $y = x^2 - 2x - 8$
(d) $y = 4x^2 - 3x - 1$
(e) $y = x^2 - 3.464x + 3$
(f) $y = \sin x$
(g) $r = \cos\theta$
(h) $\mathrm{pH} = -\log_{10}(\mathrm{H}^+)$
(i) $x^2 + y^2 = 4$
(j) $(x - 2)^2 + (y + 4)^2 = 9$

2. Plot the following functions in plane polar coordinates from 0 to 2π (remember that in polar coordinates, negative values of r have no meaning):
(a) $r = 3$
(b) $r = \theta/36$
(c) $r = 3\sin\theta$
(d) $r = 3\cos\theta$
(e) $r = 3\sin\theta\cos\theta$
(f) $r = 3\cos^2\theta - 1$

3. Plot the following functions in Cartesian coordinates:
(a) $y = 4$
(b) $y = 4x - 3$
(c) $s = 3t^2$
(d) $\psi = \sin\theta \quad (0 \text{ to } 2\pi)$
(e) $y = -2x^2 + 4x + 4$
(f) $y = (9 - x^2)^{1/2}$
(g) $y = 4e^x$
(h) $\psi = \sin\theta\cos\theta \quad (0 \text{ to } 2\pi)$

4. Plot the following functions choosing suitable coordinate axes:
(a) $E_k = \frac{1}{2}mv^2$ (constant m)
(b) $V = -e^2/r$ (e is a constant, not the exponential)
(c) $F = e^2/r^2$ (e is a constant, not the exponential)
(d) $[A] = [A]_0\, e^{-kt}$ ($[A]_0$ and k are constants; e is exponential)
(e) $1/[A] = kt + C$ (k and C constants)
(f) $k_r = Ae^{-E/RT}$ (E, R, and A constants)

5. Plot the functions in Problem 4 choosing coordinates so that a straight line results.

6. Evaluate $[f(x + h) - f(x)]/h$ for the following:
(a) $f(x) = 1/x$
(b) $f(x) = 1/x^2$
(c) $f(x) = 4x^2 - 4$
(d) $f(x) = 1/(1 + x)$

7. Radioactive decay is a first-order process in which the concentration of the radioactive material C is related to time t by the equation

$$C = C_0 e^{-kt}$$

where C_0 and k are constants (e is the exponential). Given the following data, determine the values of C_0 and k by plotting the data in such a way that a straight line results.

t	0	10	20	30	40	50	60	70
C	10	8.7	7.6	6.5	5.8	5.0	4.4	3.8

8. Using the graphical method, determine the roots of the following equations:

(a) $y = x^3 + x^2 - 2x - 1$

(b) $y = x^4 - 3x^2 + 1$

(c) $y = x^4 + x^3 - 2.111x^2 - 2x + 0.111$

(d) $y = x^6 - 6x^4 + 9x^2 - 4$

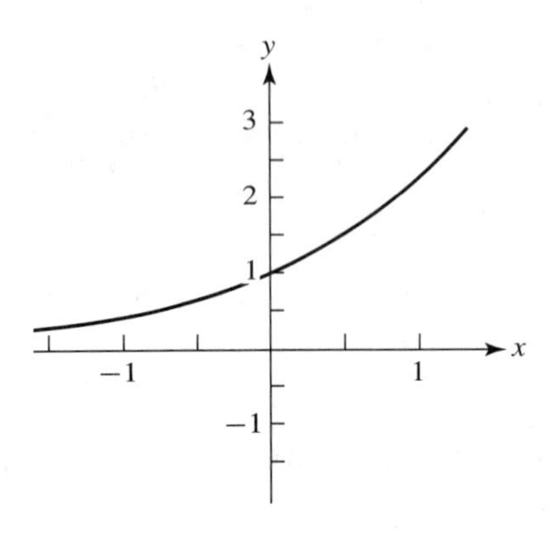

3

Logarithms

3-1 INTRODUCTION

In the previous chapter we defined a logarithm as the exponent or power x to which a number a is raised to give the number y, where $a > 0$. That is, if $a^x = y$, then $\log_a y = x$ (read "log of y to the base a"). Since the equations

$$a^x = y \quad \text{and} \quad \log_a y = x$$

are equivalent, we can use this fact to derive several useful properties of logarithms.

3-2 GENERAL PROPERTIES OF LOGARITHMS

PRODUCT RULE. *The logarithm of the product of two numbers m and n is equal to the sum of the logarithms of m and n.*

$$\log_a mn = \log_a m + \log_a n \tag{3-1}$$

Proof. Let $m = a^x$ and $n = a^y$. Then, $x = \log_a m$ and $y = \log_a n$. Now

$$mn = a^x a^y = a^{(x+y)} \tag{3-2}$$

Taking the logarithm to the base a of Equation (3-2),

$$\log_a mn = \log_a a^{(x+y)} = x + y = \log_a m + \log_a n$$

QUOTIENT RULE. *The logarithm of the quotient of the two numbers m and n is equal to the difference of the logarithms of m and n.*

$$\log_a \left(\frac{m}{n}\right) = \log_a m - \log_a n \tag{3-3}$$

Proof. Let $m = a^x$ and $n = a^y$. Then, $x = \log_a m$ and $y = \log_a n$. Now

$$\frac{m}{n} = \frac{a^x}{a^y} = a^{(x-y)} \tag{3-4}$$

$$\log_a \left(\frac{m}{n}\right) = \log_a a^{(x-y)} = x - y = \log_a m - \log_a n$$

POWER RULE. *The logarithm of m raised to the power n is equal to n multiplied by the logarithm of m.*

$$\log_a (m)^n = n \log_a m \tag{3-5}$$

Proof. Let $m = a^x$. Then, $x = \log_a m$. Now

$$m^n = (a^x)^n = a^{xn} \tag{3-6}$$

Taking the logarithm to the base a of Equation (3-6) gives

$$\log_a (m)^n = \log_a a^{nx} = nx = n \log_a m$$

3-3 COMMON LOGARITHMS

In the previous examples we did not specify any particular value for the base a; that is, the above rules hold for any value of a. In numerical calculations, however, we find that it is convenient to use logarithms to the base 10, since they are directly related to our decimal system of expressing numbers and also are linked to what normally we refer to as *scientific notation*, in which we express numbers in terms of powers of 10 (e.g., 6.022×10^{23}). Such logarithms are called *common logarithms*, and are written simply as log y. The relationship between exponents of the number 10 and common logarithms can be seen in Table 3-1.

TABLE 3-1 RELATIONSHIP BETWEEN $y = 10^x$ AND $\log_{10} y = x$

$10^0 = 1$	$\log 1 = 0$	$10^{0.3010} = 2$	$\log 2 = 0.3010$
$10^1 = 10$	$\log 10 = 1$	$10^{0.4771} = 3$	$\log 3 = 0.4771$
$10^2 = 100$	$\log 100 = 2$	$10^{0.9542} = 9$	$\log 9 = 0.9542$
$10^3 = 1000$	$\log 1000 = 3$	$10^{1.3010} = 20$	$\log 20 = 1.3010$

In general, a logarithm is composed of two parts: a *mantissa*, a positive number that determines the exact value of the number from 1 to 9.999 . . . , and a *characteristic* (multiplier) that can be positive or negative and determines where the decimal point is placed in the number. It is equivalent to expressing all numbers in scientific notation, for example, 12200 as 1.22×10^4. The number 1.22 is equivalent to the mantissa of the logarithm, and 10^4, which tells us where the decimal point is placed, is equivalent to the characteristic of the logarithm. In fact, if we determine the logarithm of 12200, we see that it is equal to the logarithm of 1.22 plus the logarithm of 10^4

$$\log (12200) = \log (1.22) + \log (10^4) = 0.0864 + 4 = 4.0864$$

Here, 0.0864 is the mantissa and 4 is the characteristic. It is important to note that the number of significant figures in a number is related to the mantissa of the logarithm and not the characteristic. The number 12200 has three significant figures, and it is the mantissa that reflects that fact. The characteristic 4 in the logarithm tells us only where the decimal point is placed.

A negative characteristic designates a number lying in a range $0 < N < 1$. To emphasize the fact that this logarithm is made up of a negative characteristic and a positive mantissa (mantissas are never negative), the minus sign is placed above the characteristic. Thus, $\log 0.020 = \log (2.0 \times 10^{-2})$ is expressed as $\bar{2}.3010$. Such a logarithm is called a *heterogeneous* logarithm. It is possible to combine the negative characteristic with the positive mantissa to form a *homogeneous* logarithm. Calculators and computers automatically do this. When this is done, the negative sign is placed in front of the logarithm, $\bar{2}.3010 = -1.6990$. The importance of the homogeneous logarithm to physical chemistry must be emphasized. All logarithmic data and certain physical quantities, such as pH, are expressed in homogeneous form. Likewise, all graphical axes involving logarithms are expressed in homogeneous form, since logarithms expressed in heterogeneous form could never be scaled conveniently on a graphical axis.

Before the age of hand calculators, common logarithms were used quite extensively to do many types of calculations which today would seem rather trivial, such as determining the roots of a number. (Find the square root or the fifth root of 4.669 without the use of a calculator!) In fact, determining the logarithm of a number itself required wading through tables of numbers. But the calculator has changed all that, and today we easily can determine the common log of any number by simply pressing a key on the calculator. We find, however, that while today it may not be necessary to use logarithms to multiply numbers together or to find the roots of numbers, they are still important, since a number of important chemical concepts, such as pH and optical absorbance, are defined in terms of the common or base-10 log. So it is important for students of physical chemistry to become familiar with the log key (and the inverse or antilog key) on their calculators.

3-4 NATURAL LOGARITHMS

In Chapter 2 we introduced a function $f(x) = e^x$ as being an exponential function particularly important to the study of physical chemistry. Logarithms taken to the base e are known as *natural logarithms* and are designated $\ln y = x$. Before going into the physical significance of the natural logarithm, it might be useful to consider the relationship between natural and common logarithms. Consider the equation

$$y = e^x \tag{3-7}$$

Taking the logarithm to the base 10 of Equation (3-7) gives

$$\log y = \log e^x = x \log e \tag{3-8}$$

However, $x = \ln y$. Substituting this into Equation (3-8) gives

$$\log y = \ln y \log e = \ln y \log (2.718)$$

But $\log (2.718) = 0.4343$. Therefore,

$$\log y = 0.4343 \ln y$$

or

$$\ln y = 2.303 \log y \tag{3-9}$$

The physical significance of the natural logarithm can best be explained with the following example. The fractional change in any variable x of a system can be written $\Delta x/x$, where Δx represents some finite change in x. Consider a variable of a system y that changes to a new value $y + \Delta y$. The fractional change in the variable is $\Delta y/y$. If the change in the variable is small, then the change in the natural logarithm of the variable is

$$\begin{aligned}\Delta \ln y &= \ln (y + \Delta y) - \ln y \\ &= \ln\left(\frac{y + \Delta y}{y}\right) = \ln\left(1 + \frac{\Delta y}{y}\right)\end{aligned}$$

Dividing both sides of the equation by Δy gives

$$\begin{aligned}\frac{\Delta \ln y}{\Delta y} &= \frac{1}{\Delta y} \ln\left(1 + \frac{\Delta y}{y}\right) \\ &= \ln\left(1 + \frac{\Delta y}{y}\right)^{1/\Delta y} \\ &= \frac{1}{y} \ln\left(1 + \frac{\Delta y}{y}\right)^{y/\Delta y}\end{aligned}$$

Remember from Chapter 2, however, that the exponential e is defined as $\lim_{x \to 0}(1+x)^{1/x}$. If we let $x = \Delta y / y$, we see that

$$\lim_{\Delta y \to 0}\left(1 + \frac{\Delta y}{y}\right)^{y/\Delta y} = \lim_{x \to 0}(1+x)^{1/x} = e$$
$$\frac{\Delta \ln y}{\Delta y} = \frac{1}{y}\ln e = \frac{1}{y} \qquad (3\text{-}10)$$

since $\ln e = 1$. Rearranging Equation (3-10) gives

$$\frac{\Delta y}{y} = \Delta \ln y \qquad (3\text{-}11)$$

In general, then, we can state that in the limit that the change in any variable x is vanishingly small, the fractional change in the variable $\Delta x/x$ is equal to the change in the natural logarithm of the variable.

$$\lim_{\Delta x \to 0}\left(\frac{\Delta x}{x}\right) = \Delta \ln x \qquad (3\text{-}12)$$

Equation (3-12) can be used to show another very important property of natural logarithms. When the change in any variable is very small and the rate of change of the natural logarithm of the variable x is constant, then the rate of change of the variable itself is directly proportional to the variable itself. This type of change is called an *exponential increase* or *decrease* and is typical of what normally are called *first-order rate processes*. The rate of change of the natural logarithm of the variable can be expressed as $\Delta \ln x / \Delta t$, where t is time. Thus, we can write

$$\frac{\Delta \ln x}{\Delta t} = k$$

Substituting for Equation (3-12), we have

$$\lim_{\Delta x \to 0} \frac{\Delta x}{x\,\Delta t} = k$$
$$\lim_{\Delta x \to 0} \frac{\Delta x}{\Delta t} - kx$$

where $\Delta x/\Delta t$ represents the rate of change of the variable.

SUGGESTED READING

1. Bradley, Gerald L., and Smith, Karl J., *Calculus*, Prentice-Hall, Inc., Upper Saddle River, NJ, 1995.
2. Sullivan, Michael, *College Algebra,* 4th ed., Prentice-Hall, Inc., Upper Saddle River, NJ, 1996.

3. VARBERG, DALE, and PURCELL, EDWIN J., *Calculus*, 7th ed., Prentice-Hall, Inc., Upper Saddle River, NJ, 1997.

PROBLEMS

1. The apparent pH of an aqueous solution is defined by the equation $\text{pH} = -\log_{10}(\text{H}^+)$. Find the apparent pH of the following solutions:

(a) $(\text{H}^+) = 1.00 \times 10^{-7}M$
(b) $(\text{H}^+) = 0.111M$
(c) $(\text{H}^+) = 9.433 \times 10^{-9}M$
(d) $(\text{H}^+) = 1.416M$
(e) $(\text{H}^+) = 5.44 \times 10^{-2}M$
(f) $(\text{H}^+) = 12.0M$

2. Given the following values for the apparent pH, find (H^+) in the following solutions:

(a) $\text{pH} = 0$
(b) $\text{pH} = 2.447$
(c) $\text{pH} = 5.893$
(d) $\text{pH} = 7.555$
(e) $\text{pH} = -0.772$
(f) $\text{pH} = 12.115$

3. Find the pH of a solution of HCl in which the HCl concentration is $1.00 \times 10^{-8}M$.

4. The work done in the isothermal, reversible expansion or compression of an ideal gas from volume V_1 to volume V_2 is given by the equation

$$w = -nRT \ln \frac{V_2}{V_1}$$

where n is the number of moles of the gas, R is the gas constant $= 8.314$ J/mol · K, and T is the absolute temperature. Find the work done in the isothermal, reversible expansion of 1.00 mole of an ideal gas at 300K from a volume of 3.00 liters to a volume of 10.00 liters.

5. The entropy change associated with the expansion or compression of an ideal gas is given by the equation

$$\Delta S = nC_v \ln \frac{T_2}{T_1} + nR \ln \frac{V_2}{V_1}$$

where n is the number of moles of the gas, C_v is the molar heat capacity at constant volume, T is the absolute temperature, and V is the volume. Find the change in entropy attending the expansion of 1.00 mole of an ideal gas from 1.00 liter to 5.00 liters, if the temperature drops from 300K to 284K. Take $C_v = \frac{3}{2}R$ and $R = 8.314$ J/mol · K.

6. It is well known that the change in entropy for an adiabatic reversible expansion of an ideal gas is equal to zero. Using the equation given in Problem 5, find the final temperature when an ideal gas at 300K expands adiabatically from 1.00 liter to 5.00 liters. Take $C_v = \frac{3}{2}R$ and $R = 8.314$ J/mol · K.

7. Radioactive decay is a first-order kinetics process which follows the integrated rate equation

$$\ln \frac{(A)}{(A)_0} = -kt$$

where (A) is the concentration of A at time t, $(A)_0$ is the concentration of A at $t = 0$ (the initial concentration of A), and k is a constant, called the rate constant. The fraction of ^{14}C found in a sample of wood ash from an archeological dig was found to be 0.664. How old is the wood ash, given that $k = 1.24 \times 10^{-4}\ \text{yr}^{-1}$ for the isotope ^{14}C.

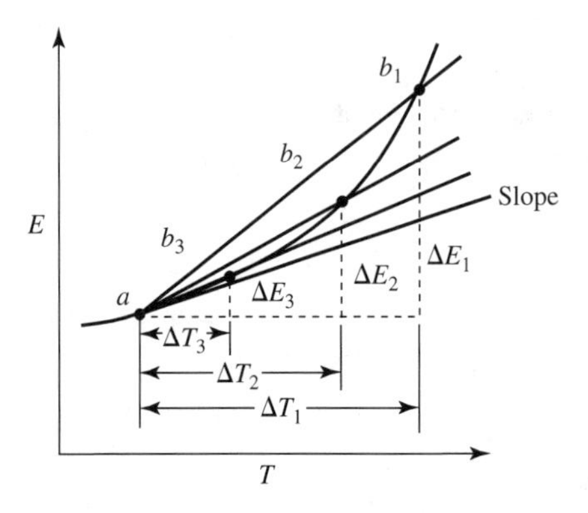

4

Differential Calculus

4-1 INTRODUCTION

Physical chemistry is concerned to a great extent with the effect that a change in one variable of a system will have on the other variables of the system. For example, how will a change in the pressure or temperature of a system affect its volume or energy? Differential calculus is the mathematics of incremental changes. It is based primarily on the mathematical concept known as the *derivative*. The derivative of a variable y with respect to a variable x, where y must be a function of x, is defined as

$$\frac{dy}{dx} = \lim_{\Delta x \to 0} \frac{\Delta y}{\Delta x} \tag{4-1}$$

where Δy and Δx denote changes in the variables y and x, respectively. Thus, the derivative of y with respect to x is simply the change in $f(x)$ with respect to the change in x, when the change in x becomes vanishingly small. If y is not a function of x, then the derivative does not exist (i.e., is equal to zero).

It is important to emphasize that, while mathematically it might be a more or less straightforward procedure to take the derivative of a function once the equation describing the functional dependence is known, it is the job of the scientist to determine how one variable of a system depends on other variables and to find the equation relating them. Some scientists are exceptionally good at doing this and win Nobel prizes, and the rest of us keep on trying. This is why it is so important, not only to understand the mathematics, but also to learn and understand the science. For example, students insist on describing the isothermal (constant temperature) expansion of a gas held by piston-cylinder arrangement (see Fig. 4-3) against a constant external pressure as an isobaric (constant pressure) process. If a gas, ideal or otherwise, expands at constant temperature, its pressure has to change (Boyle's law for an ideal gas!). There is no functional dependence between the external pressure on the gas (part of the surroundings) and the volume of the gas (part of the system). No derivative exists. Only in the very special case of the reversible expansion or compression of a gas, which we will discuss in a subsequent section, can they be related, and then only indirectly.

The derivative of a function may be taken more than once, giving rise to second, third, and higher derivatives, denoted d^2y/dx^2, d^3y/dx^3, and so on, respectively. Note that the second derivative, for example, is the first derivative of the first derivative,

$$\frac{d}{dx}\left(\frac{dy}{dx}\right) = \frac{d^2y}{dx^2}$$

the third derivative is the first derivative of the second derivative,

$$\frac{d}{dx}\left(\frac{d^2y}{dx^2}\right) = \frac{d^3y}{dx^3}$$

and so on. The process of taking derivatives of functions is called *differentiation*.

There are many uses for differential calculus in physical chemistry; however, before going into these, let us first review the mechanics of differentiation. The functional dependence of the variables of a system may appear in many different forms: as first- or second-degree equations, as trigonometric functions, as logarithms or exponential functions. For this reason, consider the derivatives of these types of functions that are used extensively in physical chemistry. Also included in the list below are rules for differentiating sums, products, and quotients. In some cases, examples are given in order to illustrate the application to physicochemical equations.

4-2 FUNCTIONS OF SINGLE VARIABLES

1. $\frac{d}{dx}(c) = 0,$ where c is any quantity not dependent on x.

2. $\frac{d}{dx}(cx) = c,$ where c is any quantity not dependent on x.

Examples:

(a) $P = kT; \quad \frac{dP}{dT} = k = \frac{P}{T}$

(b) $y = mx + b; \quad \frac{dy}{dx} = m$

3. $\frac{d}{dx}(x^n) = nx^{n-1},$ where n is any real number.

Examples:

(a) $P = \frac{k}{V} = kV^{-1}; \quad \frac{dP}{dV} = (-1)kV^{-2} = -\frac{k}{V^2} = -\frac{P}{V}$

$$\frac{d^2P}{dV^2} = (-2)(-1)kV^{-3} = \frac{2k}{V^3}$$

(b) $V = \frac{4}{3}\pi r^3; \quad \frac{dV}{dr} = (3)\left(\frac{4}{3}\right)\pi r^2 = 4\pi r^2$

(c) $E_k = \frac{1}{2}mv^2; \quad \frac{dE_k}{dv} = (2)\left(\frac{1}{2}\right)mv = mv$

(d) $\ln P = \frac{-\Delta H}{RT} + c$, where ΔH, R, and c are constants.

$$\frac{d(\ln P)}{dT} = \frac{\Delta H}{RT^2}$$

(e) $r = \sqrt{A} = (A)^{1/2}; \quad \frac{dr}{dA} = \left(\frac{1}{2}\right)(A)^{-1/2}$

4. $\frac{d}{dx}(\sin ax) = a\cos ax$, where a is a constant.

Examples:

(a) $\psi = A\sin\frac{n\pi x}{a}; \quad \frac{d\psi}{dx} = A\left(\frac{n\pi}{a}\right)\cos\frac{n\pi x}{a}$

(b) $y = A\sin(2\pi\nu t)$, where π and ν are constants.

$$\frac{dy}{dt} = A(2\pi\nu)\cos(2\pi\nu t)$$

5. $\frac{d}{dx}(\cos ax) = -a\sin ax$, where a is a constant.

Examples:

(a) $\frac{d\psi}{dx} = A\left(\frac{n\pi}{a}\right)\cos\frac{n\pi x}{a}; \quad \frac{d^2\psi}{dx^2} = -A\left(\frac{n^2\pi^2}{a^2}\right)\sin\frac{n\pi x}{a}$

(b) $y = A\cos(2\pi\nu t); \quad \frac{dy}{dt} = -A(2\pi\nu)\sin(2\pi\nu t)$

6. $\frac{d}{dx}(\tan x) = \sec^2 x$

7. $\frac{d}{dx}(e^{ax}) = ae^{ax}$, where a is a constant.

Examples:

(a) $\Phi = A\, e^{im\phi}$, where A, i, and m are constants. The constant $i = \sqrt{-1}$.

$$\frac{d\Phi}{d\phi} = imA\, e^{im\phi}$$

$$\frac{d^2\Phi}{d\phi^2} = -m^2 A\, e^{im\phi} \quad \text{or} \quad \frac{d^2\Phi}{d\phi^2} = -m^2\Phi$$

(b) $(A) = (A)_0\, e^{-kt}$, where $(A)_0$ and k are constants.

$$\frac{d(A)}{dt} = -k(A)_0\, e^{-kt}$$

8. $\dfrac{d}{dx}(\ln x) = \dfrac{1}{x}$

9. $\dfrac{d}{dx}\left[f(g(x))\right] = \dfrac{df}{dg} \cdot \dfrac{dg}{dx}$

Examples:

(a) $\Phi = 3\cos^2\theta - 1$.

Let $u = \cos\theta$; $\dfrac{du}{d\theta} = -\sin\theta$.

$$\Phi = 3u^2 - 1,\ \frac{d\Phi}{du} = 6u$$

$$\frac{d\Phi}{d\theta} = \frac{d\Phi}{du} \cdot \frac{du}{d\theta} = -6u\sin\theta = -6\cos\theta\sin\theta$$

(b) $y = \dfrac{1}{\sqrt{1-x^2}} = (1-x^2)^{-1/2}$

Let $u = (1-x^2)$; $\dfrac{du}{dx} = -2x$, and $y = u^{-1/2}$.

$$\frac{dy}{du} = -\frac{1}{2}u^{-3/2};\ \frac{dy}{dx} = \frac{dy}{du} \cdot \frac{du}{dx}$$

$$= -\frac{1}{2}(1-x^2)^{-3/2}(-2x) = \frac{x}{(1-x^2)^{3/2}}$$

(c) $n = n_0\, e^{-E/kT}$, where n_0, E, and k are constants.

Let $u = E/kT$; $\dfrac{du}{dT} = -\dfrac{E}{kT^2}$.

$$n = n_0\, e^{-u};\ \frac{dn}{du} = -n_0\, e^{-u}$$

$$\frac{dn}{dT} = \frac{dn}{du} \cdot \frac{du}{dT} = -n_0\, e^{-E/kT}\left(\frac{-E}{kT^2}\right) = n_0\left(\frac{E}{kT^2}\right)e^{-E/kT}$$

(d) $y = Ae^{-ax^2}$, where A and a are constants.

Let $u = x^2$; $\frac{du}{dx} = 2x$.

$$y = Ae^{-au}; \quad \frac{dy}{du} = -aAe^{-au}$$

$$\frac{dy}{dx} = \frac{dy}{du} \cdot \frac{du}{dx} = -aAe^{-ax^2}(2x) = -2ax\ Ae^{-ax^2}$$

10. $\frac{d}{dx}\big(f(x) + g(x)\big) = \frac{df}{dx} + \frac{dg}{dx}$

Example: $\ln P = -\frac{a}{T} + b \ln T + c$, where a, b, and c are constants.

$$\frac{d \ln P}{dT} = \frac{a}{T^2} + \frac{b}{T}$$

11. $\frac{d}{dx}\big(f(x) \cdot g(x)\big) = f(x)\frac{dg}{dx} + g(x)\frac{df}{dx}$

Examples:

(a) $y = \sin x e^{mx}$, where m is a constant.
Let $f(x) = \sin x$ and $g(x) = e^{mx}$.

$$\frac{df}{dx} = \cos x \quad \text{and} \quad \frac{dg}{dx} = me^{mx}$$

$$\frac{dy}{dx} = m \sin x e^{mx} + \cos x e^{mx}$$

(b) $F = -\eta\ 2\pi r L \frac{dy}{dr}$, where η, π, and L are constants.

Let $f(r) = -\eta\ 2\pi r L$ and $g(r) = \frac{dy}{dr}$.

$$\frac{df}{dr} = -\eta\ 2\pi L \quad \text{and} \quad \frac{dg}{dr} = \frac{d^2y}{dr^2}$$

$$\frac{dy}{dr} = -\eta\ 2\pi r L \frac{d^2y}{dr^2} - \eta\ 2\pi L \frac{dy}{dr}$$

(c) $E = kT^2 \frac{d}{dT}(\ln q)$, where k is a constant.

Let $f(T) = kT^2$ and $g(T) = \frac{d}{dT}(\ln q)$.

$$\frac{df}{dT} = 2kT \quad \text{and} \quad \frac{dg}{dT} = \frac{d^2}{dT^2}(\ln q)$$
$$\frac{dE}{dT} = kT^2\frac{d^2}{dT^2}(\ln q) + 2kT\frac{d}{dT}(\ln q)$$

(d) $\psi(x) = e^{-x^2/2}y(x)$

Let $u = -\dfrac{x^2}{2}$; $\dfrac{du}{dx} = -x$; $\dfrac{d^2u}{dx^2} = -1$. Therefore, $\psi(x) = e^u y(x)$.

$$\frac{d\psi}{dx} = e^u\frac{dy}{dx} + ye^u\frac{du}{dx}$$
$$\frac{d^2\psi}{dx^2} = e^u\frac{d^2y}{dx^2} + \frac{dy}{dx}e^u\frac{du}{dx} + y\left[e^u\frac{d^2u}{dx^2} + \frac{du}{dx}e^u\frac{du}{dx}\right] + e^u\frac{du}{dx}\frac{dy}{dx}$$
$$\frac{d^2\psi}{dx^2} = e^{-x^2/2}\frac{d^2y}{dx^2} - xe^{-x^2/2}\frac{dy}{dx} - ye^{-x^2/2} + x^2ye^{-x^2/2} - xe^{-x^2/2}\frac{dy}{dx}$$
$$= e^{-x^2/2}\frac{d^2y}{dx^2} - 2xe^{-x^2/2}\frac{dy}{dx} + x^2ye^{-x^2/2} - ye^{-x^2/2}$$

12. $$\frac{d}{dx}\left(\frac{f(x)}{g(x)}\right) = \frac{g(x)\frac{df}{dx} - f(x)\frac{dg}{dx}}{\left(g(x)\right)^2}$$

Example: $y = \tan x = \dfrac{\sin x}{\cos x}$

Let $f(x) = \sin x$ and $g(x) = \cos x$.

$$\frac{df}{dx} = \cos x \quad \text{and} \quad \frac{dg}{dx} = -\sin x$$
$$\frac{dy}{dx} = \frac{\cos x(\cos x) + (\sin x)(\sin x)}{\cos^2 x} = \frac{1}{\cos^2 x} = \sec^2 x$$

4-3 FUNCTIONS OF SEVERAL VARIABLES—PARTIAL DERIVATIVES

In all the cases given above, the functions that were differentiated contained only one independent variable. Most physicochemical systems, however, normally contain more than one independent variable. For example, the pressure of an ideal gas is simultaneously a function of the temperature of the gas and the volume of the gas. This can be expressed in the form of an equation of state for the gas

$$P = f(T, V) = \frac{RT}{V} \tag{4-2}$$

where R is a constant. Since both variables can change, let us consider two ways to treat this situation. In this section we shall consider the case where only one of the independent variables changes while the other remains constant. The derivative of P with respect to only one of the variables T or V while the other remains constant is called a *partial derivative* and is designated by the symbol ∂. The partial derivative of P with respect to V at constant T can be defined as

$$\left(\frac{\partial P}{\partial V}\right)_T = \lim_{\Delta V \to 0} \frac{f(T, V + \Delta V) - f(T, V)}{\Delta V} \tag{4-3}$$

and the partial derivative of P with respect to T at constant V can be defined as

$$\left(\frac{\partial P}{\partial T}\right)_V = \lim_{\Delta T \to 0} \frac{f(T + \Delta T, V) - f(T, V)}{\Delta T} \tag{4-4}$$

The small subscripts T and V in the expressions $(\ )_T$ and $(\ )_V$ indicate which variables are to be held constant.

The rules for partial differentiation are the same as those for ordinary differentiation (found in Section 4-2), with the addition that the variables held constant are treated the same as the other constants in the equation. Hence, at constant T

$$P = \frac{RT}{V}; \left(\frac{\partial P}{\partial V}\right)_T = \frac{-RT}{V^2} \tag{4-5}$$

and at constant V

$$P = \frac{RT}{V}; \left(\frac{\partial P}{\partial T}\right)_V = \frac{R}{V} \tag{4-6}$$

Functions of two or more variables can be differentiated partially more than once with respect to either variable while holding the other constant to yield second and higher derivatives. For example,

$$\left(\frac{\partial}{\partial T}\left(\frac{\partial P}{\partial T}\right)_V\right)_V = \left(\frac{\partial^2 P}{\partial T^2}\right)_V \tag{4-7}$$

and

$$\left(\frac{\partial}{\partial V}\left(\frac{\partial P}{\partial T}\right)_V\right)_T = \left(\frac{\partial^2 P}{\partial V \partial T}\right) \tag{4-8}$$

Equation (4-8) is called a *mixed partial second derivative*. If a function of two or more variables and its derivatives are singlevalued and continuous, a property normally attributed to physical variables, then the mixed partial second derivatives are equal. That is,

$$\left(\frac{\partial}{\partial V}\left(\frac{\partial P}{\partial T}\right)_V\right)_T = \left(\frac{\partial}{\partial T}\left(\frac{\partial P}{\partial V}\right)_T\right)_V \tag{4-9}$$

or

$$\left(\frac{\partial^2 P}{\partial V \partial T}\right) = \left(\frac{\partial^2 P}{\partial T \partial V}\right) \tag{4-10}$$

To illustrate that partial differentiation is, in fact, no more complicated than ordinary differentiation, consider the following examples.

Examples

(a) $d = \frac{m}{V}; \left(\frac{\partial d}{\partial m}\right)_V = \frac{1}{V}; \left(\frac{\partial d}{\partial V}\right)_m = \frac{-m}{V^2}$

(b) $V = \pi r^2 h; \left(\frac{\partial V}{\partial r}\right)_h = 2\pi r h; \left(\frac{\partial V}{\partial h}\right)_r = \pi r^2$

(c) $\left(\frac{\partial E}{\partial T}\right)_V = T\left(\frac{\partial S}{\partial T}\right)_V$

Take the second derivative of E with respect to V at constant T.

$$\left(\frac{\partial}{\partial V}\left(\frac{\partial E}{\partial T}\right)_V\right)_T = T\left(\frac{\partial}{\partial V}\left(\frac{\partial S}{\partial T}\right)_V\right)_T = T\left(\frac{\partial^2 S}{\partial V \partial T}\right)$$

(d) $\left(\frac{\partial E}{\partial V}\right)_T = T\left(\frac{\partial S}{\partial V}\right)_T - P$, where $P = f(T, V)$.

Take the derivative of E with respect to T at constant V.

Since both T and $(\partial S/\partial V)$ are functions of T, the first term in the expression must be differentiated as a product.

$$\left(\frac{\partial}{\partial T}\left(\frac{\partial E}{\partial V}\right)_T\right)_V = T\left(\frac{\partial}{\partial T}\left(\frac{\partial S}{\partial V}\right)_T\right)_V + \left(\frac{\partial S}{\partial V}\right)_T\left(\frac{\partial T}{\partial T}\right) - \left(\frac{\partial P}{\partial T}\right)_V$$

Since $(\partial T/\partial T) = 1$, we can write

$$\left(\frac{\partial^2 E}{\partial T \partial V}\right) = T\left(\frac{\partial^2 S}{\partial T \partial V}\right) + \left(\frac{\partial S}{\partial V}\right)_T - \left(\frac{\partial P}{\partial T}\right)_V$$

4-4 THE TOTAL DIFFERENTIAL

We now consider the second case in which independent variables of a system may be varied, and the effect of this on the dependent variable. In the previous section we allowed only one variable to change at a time. In this section we shall consider the effect of allowing all variables to change simultaneously. Consider, again, the example $P = f(T, V)$. Let ΔP represent the change in pressure brought about by a simultaneous change in temperature and volume.

$$\Delta P = f(T + \Delta T, V + \Delta V) - f(T, V) \tag{4-11}$$

Adding and subtracting $f(T, V + \Delta V)$ to Equation (4-11) yields

$$\begin{aligned} \Delta P =& f(T + \Delta T, V + \Delta V) - f(T, V + \Delta V) \\ &+ f(T, V + \Delta V) - f(T, V) \end{aligned} \tag{4-12}$$

Multiplying the first two terms in Equation (4-12) by $\Delta T/\Delta T$ and the second two terms by $\Delta V/\Delta V$ gives

$$\begin{aligned} \Delta P =& \left[\frac{f(T + \Delta T, V + \Delta V) - f(T, V + \Delta V)}{\Delta T}\right]\Delta T \\ &+ \left[\frac{f(T, V + \Delta V) - f(T, V)}{\Delta V}\right]\Delta V \end{aligned} \tag{4-13}$$

Taking the limit as ΔT and ΔV go to zero gives

$$\begin{aligned} \lim_{\Delta P \to 0} \Delta P =& \lim_{\Delta T \to 0}\left[\frac{f(T + \Delta T, V) - f(T, V)}{\Delta T}\right]\Delta T \\ &+ \lim_{\Delta V \to 0}\left[\frac{f(T, V + \Delta V) - f(T, V)}{\Delta V}\right]\Delta V \end{aligned} \tag{4-14}$$

The terms in the brackets are just the partial derivatives $(\partial P/\partial T)_V$ and $(\partial P/\partial V)_T$. Replacing ΔP, ΔT, and ΔV with dP, dT, and dV to indicate vanishingly small changes, we can write

$$dP = \left(\frac{\partial P}{\partial T}\right)_V dT + \left(\frac{\partial P}{\partial V}\right)_T dV \tag{4-15}$$

where the expression dP represents the *total differential* of P. The terms $(\partial P/\partial T)_V\, dT$ and $(\partial P/\partial V)_T\, dV$ are called *partial differentials*. The combination of the partial differentials yields the *total differential* of the function. In general, then, if a variable $u = f(x_1, x_2, x_3, \ldots)$, where $x_1, x_2, x_3, \ldots$ are independent variables,[1] then

$$du = \left(\frac{\partial u}{\partial x_1}\right)_{x_2, x_3, \ldots} dx_1 + \left(\frac{\partial u}{\partial x_2}\right)_{x_1, x_3, \ldots} dx_2 + \left(\frac{\partial u}{\partial x_3}\right)_{x_1, x_2, \ldots} dx_3 + \cdots \tag{4-16}$$

To illustrate the physical significance of Equation (4-16), consider the following example. The volume of a cylinder is a function of both the radius of the cylinder r and the height of the cylinder h and is given by the equation

$$V = f(r, h) = \pi r^2 h \tag{4-17}$$

[1]We find that Equation (4-16) will still hold even if all the variables are not independent.

Any change in either r or h will result in a change in V. The total differential of V, then, is

$$dV = \left(\frac{\partial V}{\partial r}\right)_h dr + \left(\frac{\partial V}{\partial h}\right)_r dh \tag{4-18}$$

Let us examine what each term in the expression means physically. For each incremental change in r, dr, or in h, dh, the volume changes. However, the manner in which the volume changes with r is different from the manner in which it changes with h. We see this by differentiating Equation (4-17) partially.

$$\left(\frac{\partial V}{\partial r}\right)_h = 2\pi r h \quad \text{and} \quad \left(\frac{\partial V}{\partial h}\right)_r = \pi r^2$$

Hence,

$$dV = 2\pi rh\, dr + \pi r^2\, dh \tag{4-19}$$

We see, then, that there are at least two ways to consider the volume of the cylinder and changes in that volume. The quantity $2\pi rh\, dr$ is the volume of a hollow cylinder of thickness dr, shown in Fig. 4-1. The total volume of the cylinder can be thought of as summing together concentric cylinders of volume $2\pi rh\, dr$ until the radius r is reached. Hence, any change in the radius of the cylinder will affect the volume by adding or subtracting concentric cylinders.

On the other hand, the quantity $\pi r^2\, dh$ represents the volume of a thin plate of thickness dh, shown in Fig. 4-1. Thus, the total volume of the cylinder also can be thought of as summing together these plates until the height h is reached. Any change in the height will affect the volume by adding or subtracting plates. The sum of these two effects results in the total change in the volume of the cylinder.

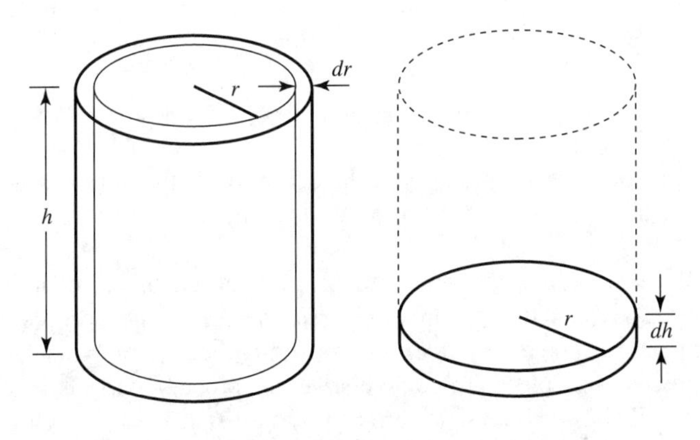

Figure 4-1 Partial differentials of a cylinder.

4-5 DERIVATIVE AS A RATIO OF INFINITESIMALLY SMALL CHANGES

In Section 4-1 we defined the derivative of $y = f(x)$ with respect to x as a ratio of the change in y to the change in x as the change in x becomes vanishingly small. We might ask at this point why such small changes are so important to the study of physical chemistry. What is the physical significance of the derivative?

To help answer these questions, consider the following example. The internal energy of a system is known to be a function of the temperature of the system, that is, $E = f(T)$. We saw in Chapter 2 that when we graph variables, such as E versus T, the relationship between changes in the two variables at some point (E, T) on the curve is given by the slope of a line drawn tangent to the curve at that point. This relationship on an E versus T curve is called the *heat capacity* of the system and is denoted by the symbol c_V. The subscript V is necessary, because we find that E also is a function of the volume V, and for this discussion we are considering V to be a constant. Thus, the heat capacity at constant volume c_V is the slope of the tangent line drawn to the E versus T curve at the point (E, T).

Consider, now, some finite change in energy $\Delta E = E_2 - E_1$ with respect to a finite change in temperature $\Delta T = T_2 - T_1$. A little experience will show us that the change in energy with respect to the change in temperature will represent the slope of the curve only when the relationship between E and T is linear, as shown in Fig. 4-2(a). Under these circumstances, it is necessary that c_V be constant with temperature. We find experimentally, however, that c_V is rarely constant with temperature, and, therefore, $\Delta E/\Delta T$ is a poor approximation to the slope of curve when E does not vary linearly with temperature, as shown in Fig. 4-2(b). Note that the ratio $\Delta E/\Delta T$, given by the line $\overline{ab}$, is quite different from the tangent to the curve at the point a, designated as the slope. We see, though, that as we allow ΔT to become smaller and smaller, the ratio $\Delta E/\Delta T$, given by lines $\overline{ab_1}$, $\overline{ab_2}$, and $\overline{ab_3}$, approaches the slope of the curve at point a, illustrated in Fig. 4-2(c). In fact, in the limit that ΔT goes to zero, the ratio $\Delta E/\Delta T$ exactly equals the slope of the curve at the point a. But this is just the definition of the derivative. That is,

$$\lim_{\Delta T \to 0} \frac{\Delta E}{\Delta T} = \frac{dE}{dT} = \text{slope of the curve}$$

Thus, we see that one useful property of the derivative is that it represents the slope of the curve (actually, the slope of a line drawn tangent to the curve) at any point along the curve.

Another example of the importance in physical chemistry of infinitesimally small changes is found in the concept of *reversibility*. A reversible process is one that, after taking place, can be reversed, exactly restoring the system to the state it was in before the process took place. To be reversible, the process must take place along a path of which all intermediate states are equilibrium states. Such a process must be defined as one in which the driving force at each step along the process is only infin-

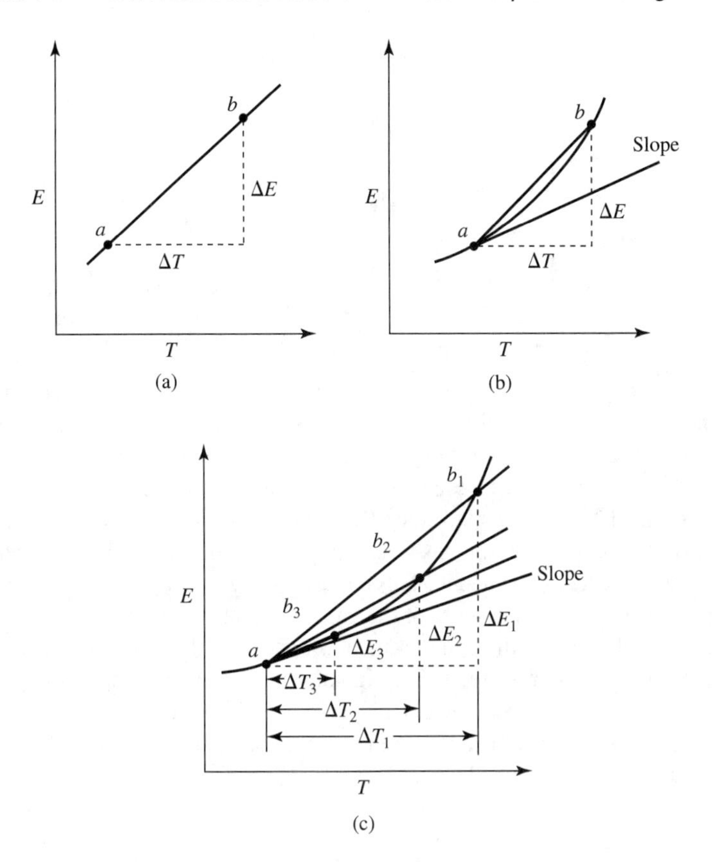

Figure 4-2 Internal energy as a function of temperature.

itesimally larger than the opposing force. If we do not define the process this way, the intermediate states will not be equilibrium states and the process will not be reversible. To illustrate this, consider the following example. Suppose we had a cylinder fitted with a frictionless piston holding one mole of an ideal gas at some initial pressure, volume, and temperature, P_1, V_1, and T_1, as shown in Fig. 4-3. Suppose that the external pressure and the gas pressure are both initially at 1 atmosphere. Next, the piston is pinned in place and the external pressure is now dropped to 0.5 atmospheres. When the pin is removed, the gas will expand suddenly, pushing the piston out to a new volume V_2 until the gas pressure drops to 0.5 atmospheres. In doing so, the gas will do a certain amount of work, $w = -P_{ext}\,\Delta V = -0.5\,\Delta V$, on the surroundings. If we assume that the cylinder is isolated from the surroundings so that no heat energy can be transferred to the gas from the surroundings, then, according to the First Law of Thermodynamics, the temperature of the gas must drop to some new value T_2 in order to account for the work done by the gas. The gas is now in a new state P_2, V_2, and T_2.

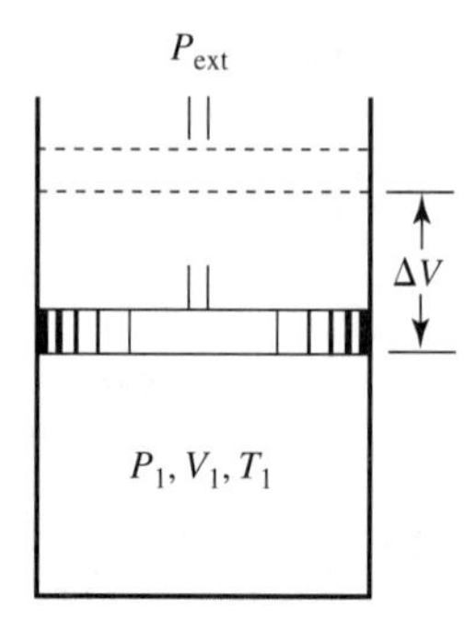

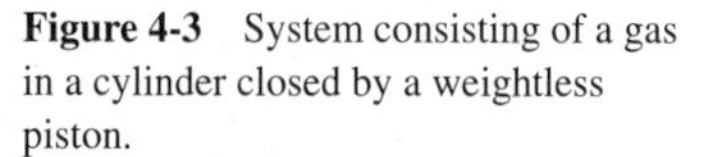
Figure 4-3 System consisting of a gas in a cylinder closed by a weightless piston.

If the above process were reversible, then we should be able to compress the gas from V_2 back to V_1, exactly restoring the system to the state that it was in before the expansion took place. The problem here is that in order to compress the gas to the volume V_1, where the gas pressure was 1 atmosphere, we must use an external pressure of at least 1 atmosphere. And since PV work depends on the external pressure, it will take twice the work on the gas to compress it to volume V_1 than the work produced by the gas when it expanded. This energy must go somewhere, and in this case it goes into raising the temperature of the gas. The final temperature of the gas when it reaches the volume V_1 will be higher than it was before the original expansion took place. The system is not restored to its original state and the process is not reversible.

Let us now repeat the above expansion, but this time let us assume that the external pressure on the gas at every point in the expansion is only infinitesimally smaller than the gas pressure. That is,

$$P_{ext} = P_{gas} - dP$$

For all practical purposes, we can consider that throughout the expansion the external pressure equals the gas pressure. As the gas expands from volume V_1 to volume V_2, the gas pressure will drop from 1 atmosphere to 0.5 atmospheres, as it did in the above example. The external pressure, however, also will do the same at every point in the expansion. Moreover, since the external pressure and the gas pressure differ by only an infinitesimal amount, the expansion should take an infinite amount of time to take place, allowing equilibrium to be established at each point in the expansion. Again, assuming that the cylinder is isolated from the surroundings, the gas temperature will drop to some new value T_2 (which is not the same as the T_2 above) to account for the work done by the gas on the surroundings as it expands. To reverse the process, we now compress the gas to its original volume by making the external pressure only infinitesimally larger than the gas pressure. That is,

$$P_{ext} = P_{gas} + dP$$

Because the external pressure at each point in the compression differs only infinitesimally from the external pressure at that point in the expansion, the work done in compressing the gas is exactly equal, but opposite, to the work done by the gas in the expansion. Upon reaching the original volume V_1, the temperature of the gas will be restored to its original temperature T_1. The system is restored to its original state and the process is truly reversible. Thus, by employing infinitesimally small changes thoughout a process, each intermediate step is allowed to reach equilibrium, and the process is reversible.

4-6 GEOMETRIC PROPERTIES OF DERIVATIVES

In the previous section we introduced the idea that in the limiting case the derivative represents an instantaneous rate of change of two variables. Hence, for example, if $y = f(x)$ is plotted on a two-dimensional Cartesian coordinate system, then dy/dx is the slope of the curve at any point (x, y) on the curve. With the exception of the function $y(x) =$ constant, functions either increase or decrease as the value of x increases. By looking at the derivative (or slope) evaluated at the point (x, y), we can determine whether the function $f(x)$ is increasing or decreasing as x increases without having to graph the function. If dy/dx is positive, then $f(x)$ increases as x increases. If dy/dx is negative, then $f(x)$ decreases as x increases.

Certain functions, such as parabolas (Fig. 4-4), or functions of higher order, such as cubic functions (Fig. 4-5), have either a maximum or a minimum value, or both. Differential calculus can be used to help us determine the point or points along the curve where maxima or minima occur. Since the slope of the curve must be zero at these points, the first derivative also must be zero. For example, the parabola shown in Fig. 4-4 is described by the equation

$$y = 2x^2 - 3x + 2$$

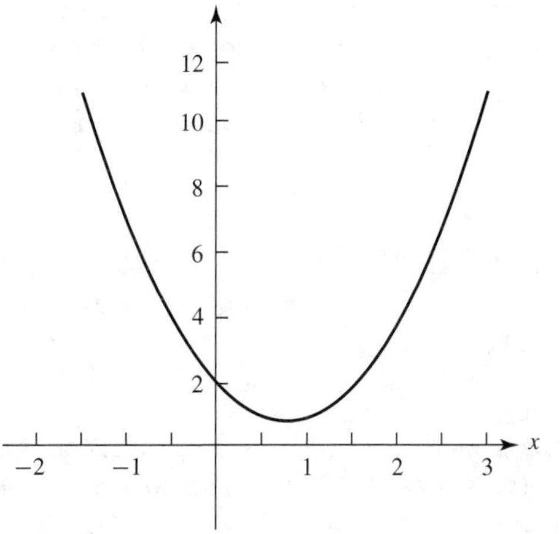

Figure 4-4 Graph of the function $y = 2x^2 - 3x + 2$.

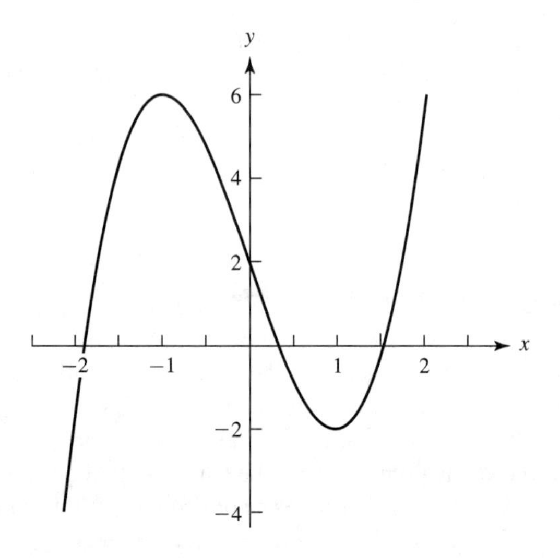

Figure 4-5 Graph of $y = 2x^3 - 6x + 2$.

Taking the first derivative gives

$$\frac{dy}{dx} = 4x - 3$$

Setting the first derivative equal to zero and solving for x, we have

$$4x - 3 = 0 \quad \text{or} \quad x = \frac{3}{4}$$

Substituting this value of x into the equation, we have $y = 0.875$, which gives the minimum point on the curve. To determine whether the curve is a maximum or a minimum at this point without having actually to graph the curve, we can substitute values for x that are both greater or smaller than $x = 0.75$ into the equation for the curve and note the behavior of y. A simpler way to test whether the function is a maximum or a minimum is to look at the second derivative of the function evaluated at the point of zero slope.

If $\dfrac{d^2y}{dx^2} < 0$, then the function is a maximum

If $\dfrac{d^2y}{dx^2} > 0$, then the function is a minimum

If $\dfrac{d^2y}{dx^2} = 0$, then a point of inflection occurs. (A point of inflection is a change from a curve that exhibits a maximum to a curve that exhibits a minimum, or vice versa.)

Consider the cubic function shown in Fig. 4-5.

$$y = 2x^3 - 6x + 2$$

Taking the first derivative and setting it equal to zero yields

$$\frac{dy}{dx} = 6x^2 - 6 = 0 \quad \text{or} \quad x^2 - 1 = 0$$

which indicates that there are two values of x for which the slope is equal to zero. Solving this equation, we see that $x = +1, -1$. Taking the second derivative of the cubic equation gives

$$\frac{d^2y}{dx^2} = 12x$$

For $x = +1$, $d^2y/dx^2 = 12$, which indicates that the curve is a minimum at this point. For $x = -1$, $d^2y/dx^2 = -12$, which indicates that the curve is a maximum at this point. Note that a point of inflection occurs at $x = 0$.

Examples

1. The total volume in milliliters of a glucose-water solution is given by the equation

$$V = 1001.93 + 111.5282m + 0.64698m^2$$

where m is the molality of the solution. The partial molar volume of glucose, $\bar{V}_{\text{glucose}}$, is the slope of a V versus m curve, $(\partial V/\partial m)$. Find the partial molar volume of glucose in a 0.100m solution of glucose in water.

Solution. Taking the derivative of the V versus m curve gives

$$\bar{V}_{\text{glucose}} = \frac{\partial V}{\partial m} = 111.5282 + 1.2940m$$

Substituting the concentration $m = 0.100$ into this equation gives the partial molar volume of glucose

$$\bar{V}_{\text{glucose}} = 111.6576 \text{ ml}$$

2. The probability of a gas molecule having a speed c lying in a range between c and $c + dc$ is given by the Maxwell distribution law for molecular speeds:

$$P_c\,dc = 4\pi \left(\frac{m}{2\pi kT}\right)^{3/2} e^{-mc^2/2kT} c^2\,dc$$

where m, k, and T are constants. Find an expression for the most probable speed.

Solution. The most probable speed occurs at the point where the probability distribution function P_c is a maximum. Thus, to determine the most probable speed, we must maximize the function P_c with respect to c.

$$P_c = 4\pi \left(\frac{m}{2\pi kT}\right)^{3/2} e^{-mc^2/2kT} c^2$$

Taking the derivative of P_c with respect to c and setting it equal to zero gives

$$\frac{dP_c}{dc} = 4\pi \left(\frac{m}{2\pi kT}\right)^{3/2} \left[e^{-mc^2/2kT}(2c) + c^2 e^{-mc^2/2kT}\left(-\frac{2mc}{2kT}\right)\right] = 0$$

We can divide through the equation by $4\pi(m/2\pi kT)^{3/2} e^{-mc^2/2kT}(2c)$, which leaves

$$1 - \frac{c^2 m}{2kT} = 0$$

$$c^2 = \frac{2kT}{m} \quad \text{or} \quad c_{mp} = \sqrt{\frac{2kT}{m}}$$

3. In the consecutive reaction $A \rightarrow B \rightarrow C$, the molar concentration of B follows the first-order rate law given by the equation

$$(B) = \frac{(A)_0 k_1}{k_2 - k_1}\left[e^{-k_1 t} - e^{-k_2 t}\right]$$

where $(A)_0$, the initial concentration of A, and the specific rate constants, k_1 and k_2, are constants. Find the value of t for which (B) is a maximum.

Solution. Taking the derivative of (B) with respect to t and setting it equal to zero gives

$$\frac{d(B)}{dt} = \frac{(A)_0 k_1}{k_2 - k_1}\left[-k_1 e^{-k_1 t} + k_2 e^{-k_2 t}\right] = 0$$

Dividing through by $(A)_0 k_1/(k_2 - k_1)$, we have

$$k_2 e^{-k_2 t} = k_1 e^{-k_1 t}$$

Taking the natural logarithm of this equation gives

$$\begin{aligned}
\ln k_2 - k_2 t &= \ln k_1 - k_1 t \\
\ln k_2 - \ln k_1 &= (k_2 - k_1)t \\
t_{\max} &= \frac{\ln(k_2/k_1)}{k_2 - k_1}
\end{aligned}$$

4-7 CONSTRAINED MAXIMA AND MINIMA

There are a number of problems in physical chemistry for which it is necessary to maximize (or minimize) a function under specific restrictive conditions. For example, suppose we wished to maximize some function $f(x, y)$ subject to the restriction that another function of x and y, $\phi(x, y)$, always equals zero. We can do this by a method known as *Lagrange's method of undetermined multipliers*. In order to maximize $f(x, y)$ by this method, consider the total differentials

$$df = \left(\frac{\partial f}{\partial x}\right)_y dx + \left(\frac{\partial f}{\partial y}\right)_x dy \tag{4-20}$$

$$d\phi = \left(\frac{\partial \phi}{\partial x}\right)_y dx + \left(\frac{\partial \phi}{\partial y}\right)_x dy = 0 \tag{4-21}$$

(Since $\phi(x, y) = 0$, $d\phi = 0$.) Equations (4-20) and (4-21) can now be combined by solving Equation (4-21) for dy and substituting this back into Equation (4-20). Hence,

$$dy = -\frac{\left(\frac{\partial \phi}{\partial x}\right)}{\left(\frac{\partial \phi}{\partial y}\right)} dx \tag{4-22}$$

and

$$df = \left[\frac{\partial f}{\partial x} - \frac{\partial f}{\partial y}\frac{\left(\frac{\partial \phi}{\partial x}\right)}{\left(\frac{\partial \phi}{\partial y}\right)}\right] dx \tag{4-23}$$

Note that this procedure effectively removes the explicit y-dependence in Equation (4-23); hence, the function f can now be treated as a function of the single variable x. Thus, the function reaches a maximum at the point where $df/dx = 0$. This gives

$$\left[\frac{\partial f}{\partial x} - \frac{\partial f}{\partial y}\frac{\left(\frac{\partial \phi}{\partial x}\right)}{\left(\frac{\partial \phi}{\partial y}\right)}\right] = 0$$

or

$$\frac{\left(\frac{\partial f}{\partial x}\right)}{\left(\frac{\partial f}{\partial y}\right)} = \frac{\left(\frac{\partial \phi}{\partial x}\right)}{\left(\frac{\partial \phi}{\partial y}\right)} \equiv \lambda \tag{4-24}$$

where λ is a constant called an *undetermined multiplier*. Rearranging Equation (4-24), we obtain two equations

$$\frac{\partial f}{\partial x} - \lambda\frac{\partial \phi}{\partial x} = 0 \quad \text{and} \quad \frac{\partial f}{\partial y} - \lambda\frac{\partial \phi}{\partial y} = 0$$

or

$$\frac{\partial}{\partial x}(f - \lambda\phi) = 0 \quad \text{and} \quad \frac{\partial}{\partial y}(f - \lambda\phi) = 0 \tag{4-25}$$

which, along with the equation $\phi(x, y) = 0$, allows us to determine the point of the maximum and λ.

We find from experience that the extension of Lagrange's method to include more than one restriction requires that there must be at least one more independent variable than there are restrictions. Thus, for more than one restriction, we have

$$\begin{aligned} F(x, y, z, \ldots) = & f(x, y, z, \ldots) - \alpha u(x, y, z, \ldots) \\ & - \beta v(x, y, z, \ldots) - \cdots \end{aligned} \tag{4-26}$$

where $f(x, y, z, \ldots)$ is the function to be maximized, $u(x, y, z, \ldots)$ and $v(x, y, z, \ldots)$ are the restrictions, and α and β are undetermined multipliers. The condition for constrained maximization or minimization of $f(x, y, z, \ldots)$ is

$$\frac{\partial F}{\partial x} = 0, \quad \frac{\partial F}{\partial y} = 0, \quad \frac{\partial F}{\partial z} = 0, \ldots \tag{4-27}$$

Examples

1. Find the dimensions of a rectangular area for which the area is a maximum and the circumference is a minimum.

Solution. The area of a rectangle is $A = ab$. The circumference of a rectangle is $C = 2(a + b)$. We wish to maximize A while we minimize C. Let $\phi = C - 2(a + b) = 0$. Therefore, Equation (4-26) for this problem is

$$F(a, b) = ab - \lambda\phi = ab - \lambda C + 2\lambda(a + b)$$

and the condition for maximization of A is

$$\frac{\partial F}{\partial a} = 0 \quad \text{and} \quad \frac{\partial F}{\partial b} = 0$$

Taking the partial derivatives gives

$$\frac{\partial F}{\partial a} = b - \lambda\frac{\partial C}{\partial a} + 2\lambda = 0 \quad \text{and} \quad \frac{\partial F}{\partial b} = a - \lambda\frac{\partial C}{\partial b} + 2\lambda - 0$$

But, since C is a minimum, $\partial C/\partial a = \partial C/\partial b = 0$. Therefore, we can write

$$b + 2\lambda = 0 \text{ and } a + 2\lambda = 0$$

or

$$a = b$$

The rectangular shape with the maximum area and the minimum circumference is a square.

2. A problem in statistical mechanics requires maximizing the function

$$f(n_0, n_1, n_2, \ldots) = n \ln n - \sum_{n_i=0}^{n} n_i \ln n_i$$

subject to the conditions that

$$\sum n_i = n \quad \text{and} \quad \sum n_i E_i = E$$

Solution. Let

$$u(n_1, n_2, n_1, \ldots) = \sum n_i - n = 0$$

and

$$v(n_1, n_2, n_3, \ldots) = \sum n_i E_i - E = 0$$

Thus,

$$F(n_1, n_2, n_3, \ldots) = n \ln n - \sum_{n_i=0}^{n} n_i \ln n_i - \alpha \left(\sum n_i - n \right)$$
$$- \beta \left(\sum n_i E_i - E \right)$$

where α and β are undetermined multipliers. The condition for constrained maximization of $n \ln n - \sum_{n_i=0}^{n} n_i \ln n_i$ is

$$\frac{\partial F}{\partial n_1} = 0, \ \frac{\partial F}{\partial n_2} = 0, \ldots \quad \text{or} \quad \frac{\partial F}{\partial n_j} = 0; \ j = 0, 1, 2, 3, \ldots$$

Taking the derivative of F with respect to each n_j in the sum, and recalling that $n_j \ln n_j$ must be differentiated as a product and that n and E are constants, we have

$$\frac{\partial F}{\partial n_j} = -n_j \left(\frac{1}{n_j} \right) - \ln n_j - \alpha - \beta E_j = 0$$

$$\ln n_j = -1 - \alpha - \beta E_j$$

Taking the antilog of this equation gives

$$n_j = e^{-(1+\alpha)} e^{-\beta E_j}$$

But $n = \sum n_j$. Therefore, $n = e^{-(1+\alpha)} \sum e^{-\beta E_j}$, which gives Boltzmann's distribution equation

$$\frac{n_j}{n} = \frac{e^{-\beta E_j}}{\sum e^{-\beta E_j}}$$

The function $n \ln n - \sum_{n_i=0}^{n} n_i \ln n_i$ will be a maximum when

$$\frac{n_0}{n} = \frac{e^{-\beta E_0}}{\sum e^{-\beta E_j}}; \quad \frac{n_1}{n} = \frac{e^{-\beta E_1}}{\sum e^{-\beta E_j}}; \quad \frac{n_2}{n} = \frac{e^{-\beta E_2}}{\sum e^{-\beta E_j}}; \ldots$$

SUGGESTED READING

1. BRADLEY, GERALD L., and SMITH, KARL J., *Calculus*, Prentice-Hall, Inc., Upper Saddle River, NJ, 1995.
2. VARBERG, DALE, and PURCELL, EDWIN J., *Calculus*, 7th ed., Prentice-Hall, Inc., Upper Saddle River, NJ, 1997.

PROBLEMS

1. Differentiate the following functions (assuming the lowercase letters to be the variables and all uppercase letters to be constants):

(a) $y = 4x^3 + 7x^2 - 10x + 6$

(b) $y = \sqrt{1 - x^2}$

(c) $y = 2x^2 - 9x - 14$

(d) $r = 3 \tan 2\theta$

(e) $y = x^3 e^{2x}$

(f) $r = A \sin\theta \cos\theta$

(g) $y = x^4 \sqrt{1 - e^x}$

(h) $y = x^6(1 - e^x) \sin 4x$

(i) $y = \dfrac{x^3}{\sqrt{1 - 3x}}$

(j) $y = \ln(1 - e^x)$

(k) $w = N \ln N - n_i \ln n_i$

(l) $s = \ln t \cdot e^{-3t}$

(m) $\ln g = \dfrac{A}{t} + t \ln t$

(n) $e = \dfrac{E^2}{A}\left(z^2 - \dfrac{27}{8}z\right)$

(o) $\phi = 2A \sin\left(\dfrac{N\pi x}{L}\right)$

(p) $\ln p = \dfrac{-\Delta H}{Rt} + K$

(q) $\ln k = -\dfrac{\Delta G}{Rt}$

(r) $u = \dfrac{A}{r^{12}} - \dfrac{B}{r^6}$

(s) $d = \dfrac{M}{v}$

(t) $\phi = Ae^{-B/Rt}$

2. Evaluate the following partial derivatives:

(a) $PV = nRT$; P with respect to V

(b) $\left(P + \dfrac{n^2 a}{V^2}\right)(V - nb) = nRT$; P with respect to V

(c) $\rho = \dfrac{PM}{RT}$; ρ with respect to T

(d) $H = a + bT + cT^2 + \dfrac{d}{T}$; H with respect to T

(e) $r = \sqrt{(x^2 + y^2 + z^2)}$; r with respect to z

(f) $y = r \sin\theta \cos\phi$; y with respect to ϕ

(g) $\left(\dfrac{\partial S}{\partial T}\right)_P = \dfrac{1}{T}\left(\dfrac{\partial H}{\partial T}\right)_P$; S with respect to P at constant T

(h) $\left(\dfrac{\partial S}{\partial P}\right)_T = \dfrac{1}{T}\left[\left(\dfrac{\partial H}{\partial P}\right)_T - V\right]$; S with respect to T at constant P

Note that $(\partial H/\partial P)$ is also a function of T.

(i) $D = \sin\theta \cos\theta \cos\phi$; D with respect to θ

(j) $E = \dfrac{c_A^2 H_{AA} + c_B^2 H_{BB} + 2c_A c_B H_{AB}}{c_A^2 + c_B^2 + 2c_A c_B S_{AB}}$; E with respect to c_A

(k) $q = \sum e^{-E_i/kT}$; q with respect to E_i

(l) $q = \sum e^{-E_i/kT}$; q with respect to T

3. Determine the slope of each of the following curves at the points indicated:

(a) $y = x^2$ at $x = 3$

(b) $y = x^3 + 4x^2 - 3x + 2$ at $x = 2$

(c) $y = 4 \ln 4x$ at $x = 1$

(d) $y = x \ln x$ at $x = 5$

(e) $r = 20 \cos\theta$ at $\theta = \pi$

(f) $r = 10 \sin\theta \cos\theta$ at $\theta = \dfrac{\pi}{2}$

(g) $y = (x^2 - 5)^{1/2}$ at $x = 3$

(h) $s = \dfrac{1}{2}At^2$ at t = 20 seconds, where A (acceleration) = 9.80 m/s^2 is constant.

(i) $C_p = 25.90 + 33.00 \times 10^{-3}\, T - 30.4 \times 10^{-7}\, T^2$ at $T = 300\text{K}$

(j) $\ln P = -\Delta H/RT + B$ at $T = 300\text{K}$, where $\Delta H = 30{,}820$ J/mol, $R = 8.314$ J/mol · K, and $B = 2.83$ are constants.

(k) $(A) = (A)_0 e^{-kt}$ at $t = 5.0$ hours, where $(A)_0 = 0.01M$ and $k = 5.08 \times 10^{-2}$ hr^{-1} are constants.

4. Determine whether each of the following functions contains a maximum value, minimum value, or both. Evaluate each function having maximum or minimum values at those points. Specify any points of inflection.

(a) $y = 4x^2 - 5x + 4$

(b) $y = 2x^3 + 3x^2 - 36x + 16$

(c) $y = \sin 3x$

(d) $\psi = Ae^{mx}$, A and m constants.

(e) $U(r) = 4e\left[\left(\dfrac{\sigma}{r}\right)^{12} - \left(\dfrac{\sigma}{r}\right)^{6}\right]$, where e and σ are constants.

(f) $\psi = \dfrac{1}{2}(1 + \sin\theta) + \sqrt{2}\cos\theta$

(g) $E = \dfrac{e^2}{a}\left(z^2 - \dfrac{27}{8}z\right)$, where e and a are constants.

(h) $P_E = 2\pi^{-1/2}(kT)^{-3/2}e^{-E/kT}E^{1/2}$, where π, k, and T are constants.

(i) $P(x) = \left(\frac{2}{a}\right) \sin^2 \frac{\pi x}{a}$ between $x = 0$ and $x = a$.

(*Hint:* $\sin \frac{\pi x}{a} = 0$ only at $x = 0$ and $x = a$.)

(j) $U(r) = -N_0 A \frac{z^2}{r} + \frac{B}{r^n}$, where N_0, A, z, and B are constants.

5. The rate constant for a chemical reaction is found to vary with temperature according to the Arrhenius equation

$$k = Ae^{-E_a/RT}$$

where A, E_a, and R are constants. Find an expression that describes the change in k with respect to the change in T.

6. The density of an ideal gas is found to vary with temperature according to the equation

$$\rho = \frac{PM}{RT}$$

where P, M, and R are considered to be constants in this case. Find an expression that describes the slope of a ρ versus T curve.

7. Find the partial derivative of P with respect to T for a gas obeying van der Waals' equation

$$\left(P + \frac{n^2 a}{V^2}\right)(V - nb) = nRT$$

8. Find the partial derivative of P with respect to V for the gas in Problem 7.

9. A certain gas obeys the equation of state

$$P(V - nb) = nRT$$

where in this case n and R are constants. Determine the coefficient of expansion of this gas $\alpha = (1/V)(\partial V/\partial T)_P$.

10. The volume of an ideal gas is simultaneously a function of the pressure and temperature of the gas. Write an equation for the total differential of V. Using the ideal gas law for 1 mole of gas, $PV = RT$, evaluate the partial derivatives in the equation.

11. The vibrational potential energy of a diatomic molecule can be approximated by the Morse function

$$U(r) = A(1 - e^{-B(r-r_0)})^2$$

where A, B, and r_0 are constants. Find the value of r for which U is a minimum.

12. The equation for describing the realm of spacial possibilities for a particle confined to a one-dimensional "box" in the state $n = 1$ is

$$\psi(x) = \sqrt{\frac{2}{a}} \sin \frac{\pi x}{a}$$

where a is the length of the "box." Find the value of x for which $\psi(x)$ is a maximum.

13. The probability distribution function describing the probability of finding an electron in the 1s-orbital of the hydrogen atom a certain distance r from the nucleus is given by the equation

$$P(r) = 4\left(\frac{1}{a_0}\right)^3 e^{-2r/a_0} r^2$$

Show that this distribution function reaches a maximum at the point when $r = a_0$, the Bohr radius.

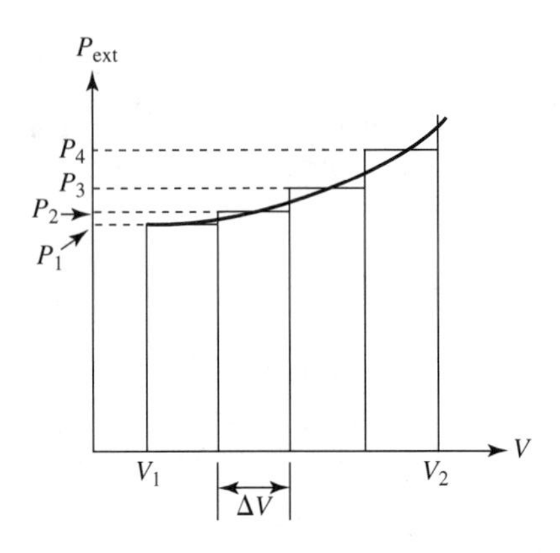

5

Integral Calculus

5-1 INTRODUCTION

There are basically two major approaches to integral calculus. One approach is to consider the integral as an antiderivative and *integration,* the process of taking integrals, as the inverse of differentiation. The other approach is to consider the integral as the sum of many similar, infinitesimal elements. The first approach allows us to mathematically generate integrals. The second approach allows us to assign a physical meaning to the integral. Introductory courses on integral calculus spend a tremendous amount of time on the first approach, teaching all the various methods for generating integrals. While this is important, and perhaps at some point should be learned, in practice it is rarely used, most of us referring to tables of integrals to do our integrating. Thus, in this text we shall emphasize using tables of integrals for performing the mechanics of integration. However, some general and special methods of integration are included, primarily because some functions are not always in a form found in the integral tables.

In the previous chapter we studied the mathematics associated with dividing a function into many small, incremental parts and determining the effects of the incremental changes on the variables of the function. In this chapter we shall consider the reverse process. Knowing the effect of the individual changes, we wish to determine the overall effect of adding together these changes such that the sum equals a finite change. Before considering the physical significance and the applications of integral calculus, let us briefly review the general and special methods of integration.

5-2 INTEGRAL AS AN ANTIDERIVATIVE

In Chapter 4 we considered the differentiation of the function $y = f(x)$, symbolized by the equation

$$\frac{dy}{dx} = \frac{df(x)}{dx} = f'(x) \tag{5-1}$$

or, in differential form,

$$dy = f'(x)\,dx \tag{5-2}$$

where $f'(x)$ denotes the first derivative of the function $f(x)$ with respect to x. In this section we shall pose the following question: What function $f(x)$, when differentiated, yields the function $f'(x)$? For example, what function $f(x)$, when differentiated, yields the function $f'(x) = 2x$? Substituting $f'(x) = 2x$ into Equation (5-2) gives

$$dy = 2x\,dx \quad \text{or} \quad \frac{dy}{dx} = 2x$$

This function, $f(x)$, for which we are looking is called the *integral* of the differential and is symbolized by the equation

$$f(x) = \int f'(x)dx \tag{5-3}$$

where the symbol $\int$ is called the *integral sign*.[1] The function that is to be integrated, $f'(x)$, is called the *integrand*.

It is not too difficult to see by inspection in this case that if $f'(x) = 2x$, then $f(x) = x^2$, since if one differentiates x^2, one obtains the derivative $f'(x) = 2x$. The term $f(x) = x^2$ is not the complete solution, however, since differentiation of the function $f(x) = x^2 + C$, where C is a constant, also will yield $f'(x) = 2x$. Hence, there is always the possibility that the integral may contain a constant, called the *constant of integration*, and this constant *always* is included as part of the answer to any integration. Thus,

$$y = \int 2x\,dx = x^2 + C$$

5-3 GENERAL METHODS OF INTEGRATION

Let us now consider several general methods of integration. Listed next are the standard integrals for most of the functions important to physical chemistry. For a complete Table of Integrals, see Appendix II.

[1]The integral sign evolved from an elongated S that originally stood for "summation."

1. $\int du(x) = u(x) + C$

The integral of the differential of a function is equal to the function itself.

2. $\int a\, du = a \int du = au + C$, where a is a constant.

Since a is a constant, it can be brought out of the integral sign.

3. $\int u^n\, du = \dfrac{u^{n+1}}{n+1} + C$, where $n \neq -1$.

Examples:

(a) $\int x^3\, dx = \dfrac{x^4}{4} + C$

(b) $\int \dfrac{\Delta H}{RT^2}\, dT$, where ΔH and R are constants.

$$\int \frac{\Delta H}{RT^2}\, dT = \frac{\Delta H}{R} \int \frac{1}{T^2}\, dT = \frac{\Delta H}{R} \int T^{-2}\, dT = -\frac{\Delta H}{R} T^{-1} + C$$
$$= -\frac{\Delta H}{RT} + C$$

4. $\int \dfrac{du}{u} = \int d \ln u = \ln u + C$

Examples:

(a) $\int \dfrac{dx}{3x} = \dfrac{1}{3} \int \dfrac{dx}{x} = \dfrac{1}{3} \ln x + C$

(b) $\int \dfrac{d(A)}{(A)} = -k \int dt$

$$\ln (A) = -kt + C$$

(c) $\int \dfrac{dP}{P} = \int \dfrac{\Delta H}{RT^2}\, dT$

$$\ln P = -\frac{\Delta H}{RT} + C$$

5. $\int [f(x) + g(x)]\, dx = \int f(x)\, dx + \int g(x)\, dx$

The integral of a sum is the sum of the integrals.

Example:

$$\int \left(a + bT + \frac{c}{T}\right) dT = \int a\, dT + \int bT\, dT + \int \frac{c}{T}\, dT$$
$$= aT + \frac{b}{2} T^2 + c \ln T + C$$

6. $\int e^{mx}\,dx = \frac{1}{m}e^{mx} + C$

7. $\int \sin kx\,dx = -\frac{1}{k}\cos kx + C,$ where k is a constant.

8. $\int \cos kx\,dx = \frac{1}{k}\sin kx + C,$ where k is a constant.

5-4 SPECIAL METHODS OF INTEGRATION

Many of the functions encountered in physical chemistry are not in one of the general forms given above. Moreover, in many cases they are not in one of the forms found in the Table of Integrals given in Appendix II. For this reason, we include several special methods of integration.

Algebraic Substitution. We find that certain mathematical functions can be transformed into one of the general forms in Section 5-3 or into one of the forms found in the Table of Integrals by some form of algebraic substitution.

Examples

(a) Evaluate $\int 2x(1-x^2)^5\,dx$.

Let us attempt to transform this integral into the form $\int u^n\,du$. Let $u = (1-x^2)$. Then $du = -2x\,dx$. Hence,

$$\int 2x(1-x^2)^5\,dx = -\int u^5\,du = -\frac{1}{6}u^6 + C = -\frac{1}{6}(1-x^2)^6 + C$$

(b) Evaluate $\int e^{-\Delta E/kT}\left(\frac{\Delta E}{kT^2}\right)\,dT$.

Let $u = -\frac{\Delta E}{kT}$. Then $du = \frac{\Delta E}{kT^2}\,dT$. Hence,

$$\int e^{-\Delta E/kT}\left(\frac{\Delta E}{kT^2}\right)\,dT = \int e^u\,du = e^u + C = e^{-\Delta EkT} + C$$

(c) Evaluate $\int \frac{dV}{V-nb}$.

Let us attempt to transform the integral into the form $\int \frac{du}{u}$.

Let $u = V - nb$. Then $du = dV$. Hence,

$$\int \frac{dV}{V-nb} = \int \frac{du}{u} = \ln u + C = \ln(V-nb) + C$$

(d) Evaluate $\int \sin^2 x \cos x\, dx$.

Let $u = \sin x$. Then $du = \cos x\, dx$. Hence,

$$\int \sin^2 x \cos x\, dx = \int u^2\, du = \frac{1}{3}u^3 + C = \frac{1}{3}\sin^3 x + C$$

Trigonometric Transformation. Many trigonometric integrals can be transformed into a proper form for integration by making some form of trigonometric transformation using trigonometric identities. For example, to evaluate the integral $\int \sin^2 x\, dx$, we must make use of the identity

$$\sin^2 x = \frac{1}{2}(1 - \cos 2x)$$

Thus,

$$\int \sin^2 x\, dx = \int \frac{1}{2}(1 - \cos 2x)\, dx = \frac{1}{2}\int dx - \frac{1}{2}\int \cos 2x\, dx$$

Integrating each term separately gives

$$\int \sin^2 x \cos x\, dx = \frac{x}{2} - \frac{1}{4}\sin 2x + C$$

Again, integration of integrals of this type are more practically done by using the Table of Integrals (Appendix II).

Example

Evaluate the integral $\int \cos^3 2x\, dx$

Integration of this function can be accomplished using Integral (88) from the Table of Integrals. Here, $a = 2$ and $b = 0$.

$$\int \cos^3 (ax + b)\, dx = \frac{1}{a}\sin(ax + b) - \frac{1}{3a}\sin^3(ax + b) + C$$

$$\int \cos^3 (2x)\, dx = \frac{1}{2}\sin(2x) - \frac{1}{6}\sin^3(2x) + C$$

Partial Fractions. Consider an integral of the type

$$\int \frac{dx}{(a - x)(b - x)}, \quad \text{where } a \text{ and } b \text{ are constants}$$

This type of integral can be transformed into simpler integrals by the following method. Let $A = (a - x)$ and $B = (b - x)$. Then

$$\frac{1}{A} - \frac{1}{B} = \frac{B}{AB} - \frac{A}{AB} = \frac{B - A}{AB}$$

Therefore,

$$\frac{1}{AB} = \frac{1}{(B - A)}\left(\frac{1}{A} - \frac{1}{B}\right)$$

$$\frac{1}{(a - x)(b - x)} = \frac{1}{(b - a)}\left(\frac{1}{(a - x)} - \frac{1}{(b - x)}\right)$$

or

$$\int \frac{dx}{(a - x)(b - x)} = \frac{1}{(b - a)}\left[\int \frac{dx}{(a - x)} - \int \frac{dx}{(b - x)}\right]$$

which can be integrated to give

$$\int \frac{dx}{(a - x)(b - x)} = \frac{1}{(b - a)}[-\ln (a - x) + \ln (b - x)] + C$$

$$= \frac{1}{(b - a)} \ln \frac{(b - x)}{(a - x)} + C$$

5-5 THE INTEGRAL AS A SUMMATION OF INFINITESIMALLY SMALL ELEMENTS

In the previous sections we have considered integration as the purely mathematical operation of finding antiderivatives. Let us now turn to the more physical aspects of integration in order to understand the physical importance of the integral.

Consider, as an example, the expansion of an ideal gas from a volume of V_1 to a volume V_2 against a constant external pressure P_{ext}, illustrated on the indicator diagram shown in Fig. 5-1. We should emphasize a very important point about this indicator diagram. The plot seen in the indicator diagram is *not* a graph of P_{ext} as a function of V. *There is no functional dependence between the external pressure on the gas and the volume of the gas.* The diagram simply shows what the external pressure is doing on one axis and what the volume is doing on the other axis. Both can vary independently. We find from physics, however, that work done by the gas does depend on both the external pressure and the volume change, and for this expansion the work is $w = -P_{ext}(V_2 - V_1)$, which we see is just the negative of the area under the P_{ext} versus V diagram.[2] By measuring this area, we can determine the work done by the gas without the need to consider the specific details of the expansion.

[2]The minus sign is necessary here because, by modern convention, work done by the system on the surroundings is defined as being negative. This contrasts with historical convention by which work done by the system on the surroundings was defined as being positive.

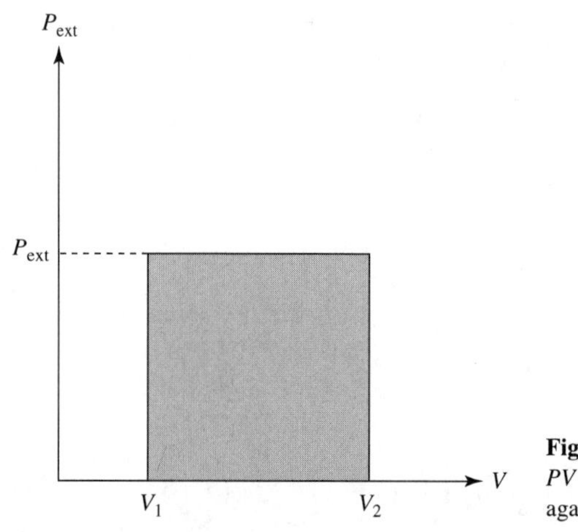

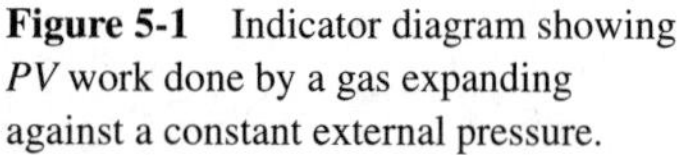
Figure 5-1 Indicator diagram showing PV work done by a gas expanding against a constant external pressure.

Consider, next, a more complicated case in which the external pressure changes, in some fashion, as the volume changes. Again, we understand that the external pressure is not changing as a function of volume. To emphasize this point, let us assume that the external pressure actually goes up as the volume changes, illustrated by the indicator diagram shown in Fig. 5-2. (A gas could be expanding against atmospheric pressure which perhaps is increasing over the period of time that the expansion takes place.) The work done in this case is still the area under a P_{ext} versus V curve. Measurement of this area, however, is much more difficult than it was when the external pressure was constant.

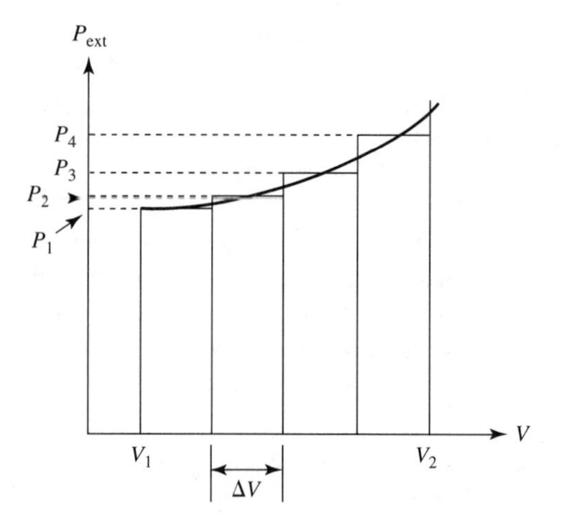

Figure 5-2 Indicator diagram showing PV work done by a gas expanding against a variable external pressure.

We can approximate the area under the curve shown in Fig. 5-2 by dividing the area into four rectangles of equal width ΔV. The approximate area under the curve, then, is just the sum of the four rectangles

$$\begin{aligned} A_{\text{approx}} &= P_1\,\Delta V + P_2\,\Delta V + P_3\,\Delta V + P_4\,\Delta V \\ &= \sum_{i=1}^{4} P_i \Delta V \end{aligned} \tag{5-4}$$

If we extend this process even further—that is, if we divide the area under the curve into more and more rectangles of smaller and smaller ΔV—the sum approaches a fixed value as N approaches infinity. Without proof, we shall define this limiting fixed value of the above summation as the true area under the curve in the interval between V_1 and V_2. Hence, we can write

$$A = \lim_{N \to \infty} \sum_{i=1}^{N} P_i\,\Delta V \tag{5-5}$$

However, since as N approaches infinity, ΔV approaches zero, we also can write

$$A = \lim_{\Delta V \to 0} \sum_{i=1}^{N} P_i\,\Delta V \tag{5-6}$$

But, by definition

$$\lim_{\Delta V \to 0} \sum_{i=1}^{N} P_i\,\Delta V = \int_{V_1}^{V_2} P_i\,dV \tag{5-7}$$

where the symbol $\int_{V_1}^{V_2}$ is read "the integral from V_1 to V_2," and V_1 and V_2 are called the *limits of integration*. Hence,

$$A = \int_{V_1}^{V_2} P_{\text{ext}}\,dV \tag{5-8}$$

The integral in Equation (5-7) is called a *definite integral*, because it has a fixed value in the interval between V_1 and V_2. We see, then, that the integral is the summation of an infinite number of infinitesimally small slices or elements of area.

5-6 LINE INTEGRALS

Having defined the integral as representing the area under a curve, we now ask whether it is possible to evaluate the integral described in Equation (5-8) using the analytical methods described in Section 5-3. For the case of P_{ext} versus V the answer is no. The reason for this is as follows. Integrals of the general type

$$A = \int_{x_1}^{x_2} y\,dx \tag{5-9}$$

are called *line integrals*, because such integrals represent the area under the specific curve (path) connecting x_1 to x_2. Such an integral can be evaluated analytically (i.e., by finding the antiderivative) only if an equation, the path, $y = f(x)$ is known, since under these circumstances the integral $\int_{x_1}^{x_2} f(x)\,dx$ contains only one variable. If y is not a function of x (as in the case of P_{ext} versus V), or if y is a function of x, but the equation relating y and x is not known or cannot be integrated (as in the case of $y = e^{-ax^2}$), or if y is a function of more than one variable that changes with x, then the line integral cannot be evaluated analytically, and one must resort to a graphical or numerical method of integration in order to evaluate the integral (see Chapter 11, Section 11-4, and Chapter 12, Section 12-6).

We sometimes can get around this problem by imposing special conditions on y. For example, the integral in Equation (5-8) can be evaluated analytically if we assume that the external pressure is a constant, since, if P_{ext} is constant, it can be brought out of the integral.

$$A = \int_{V_1}^{V_2} P_{\text{ext}}\,dV = P_{\text{ext}} \int_{V_1}^{V_2} dV = P_{\text{ext}}(V_2 - V_1)$$

$$\text{work} = -A = -P_{\text{ext}}\,(V_2 - V_1)$$

which is the area described in Fig. 5-1. Note that the definite integral is evaluated by first finding the indefinite integral, and then simply substituting the upper limit and then the lower limit into the indefinite integral, and subtracting the two. In the definite integral, the constant of integration C vanishes.

A second way to evaluate the integral in Equation (5-8) is found in the concept of reversibility. If the expansion of the gas is reversible, then for all practical purposes the external pressure is equal to the gas pressure, $P_{\text{ext}} = P_{\text{gas}}$. This gives

$$A = \int_{V_1}^{V_2} P_{\text{ext}}\,dV = \int_{V_1}^{V_2} P_{\text{gas}}(T, V)\,dV$$

Even under these circumstances, however, evaluation of the integral is still not possible, because, while P_{gas} is a function of V, it is also a function of temperature, and for each variation of temperature with volume (which describes a specific path from V_1 to V_2), the integral will have a different value. But if we further stipulate that the temperature is constant, then the integral can be evaluated. For an ideal gas under isothermal conditions, we can write

$$\begin{aligned} A &= \int_{V_1}^{V_2} P_{\text{ext}}\,dV = \int_{V_1}^{V_2} P_{\text{gas}}\,dV = \int_{V_1}^{V_2} \frac{nRT}{V}\,dV \\ &= nRT \int_{V_1}^{V_2} \frac{dV}{V} = nRT(\ln V_2 - \ln V_1) = nRT \ln \frac{V_2}{V_1} \end{aligned}$$

$$\text{work} = -A = -nRT \ln \frac{V_2}{V_1}$$

Example

The change in enthalpy as a function of temperature is given by the equation

$$\Delta H = \int_{T_1}^{T_2} C_p \, dT \qquad \text{(per mole)}$$

Find the change in enthalpy for one mole of a real gas when the temperature of the gas is increased from, say, 298.2K to 500.0K.

Solution. We first recognize that the integral above is a line integral. This integral cannot be evaluated unless C_p as a function of T is known. We could assume that C_p is a constant and evaluate the integral that way, but over such a large temperature range the approximation would be poor. Another approach would be to determine the integral numerically. (Numerical methods are covered in Chapter 11.) One analytical approach is to expand C_p as a power series in temperature. (Power series are covered in Chapter 7.) While this is not the exact functional relationship between C_p and T, we find that this approach gives good results.

$$C_p = a + bT + cT^2$$

The constants a, b, and c are known for many common gases. Substituting this into the ΔH equation gives

$$\Delta H = \int_{T_1}^{T_2} (a + bT + cT^2) \, dT$$

which now can be integrated to give

$$\Delta H = a(T_2 - T_1) + \frac{b}{2}(T_2^2 - T_1^2) + \frac{c}{3}(T_2^3 - T_1^3)$$

We see, then, that the change in enthalpy of the system is found by summing over the entire temperature range, one infinitesimal contribution $C_p \, dT$ at a time.

5-7 DOUBLE AND TRIPLE INTEGRALS

In Chapter 4 we saw that functions could be differentiated more than once. Let us consider the inverse of this process—the determination of multiple integrals. The volume of a cylinder is a function of both the radius and the height of the cylinder. That is, $V = f(r, h)$. Let us suppose that we allow the height of the cylinder, h, to change while holding the radius, r, constant. The integral from $h = 0$ to $h = h$, then, could be expressed as

$$\int_0^h f(r, h) \, dh \tag{5-10}$$

But the value of this line integral depends on the value of the radius, r, and hence the integral could be considered to be a function of r.

$$g(r) = \int_0^h f(r, h)\, dh \tag{5-11}$$

If we now allow r to vary from $r = 0$ to $r = r$ and integrate over the change, we can write

$$\int_0^r g(r)\, dr = \int_0^r \int_0^h f(r, h)\, dh\, dr \tag{5-12}$$

which is read, "the double integral of $f(r, h)$ from $h = 0$ to $h = h$ and $r = 0$ to $r = r$."

To evaluate the above double integral, we integrate $\int_0^h f(r, h)\, dh$ first while holding r a constant, which gives us $g(r)$. Then we integrate $\int_0^r g(r)\, dr$ next while holding h constant. Such a process is known as *successive partial integration*. For example, let us evaluate $\int_0^r \int_0^h 2\pi r\, dh\, dr$. First,

$$g(r) = \int_0^h 2\pi r\, dh = 2\pi r h$$

Next, we integrate

$$\int_0^r g(r)\, dr = \int_0^r 2\pi r h\, dr = \pi r^2 h$$

which one recognizes as the volume of a cylinder.

The above argument can be extended to the triple integral. For example, let us evaluate the triple integral

$$\int_0^z \int_0^y \int_0^x dx\, dy\, dz \tag{5-13}$$

First, evaluate $\int_0^x dx = x$. Substituting this back into Equation (5-13) gives

$$\int_0^z \int_0^y x\, dy\, dz$$

Next, evaluate $\int_0^y x\, dy = xy$. Substituting this back into Equation (5-13) gives

$$\int_0^z xy\, dz$$

Integrating this gives

$$\int_0^z xy\, dz = xyz$$

which is the volume of a rectangular box x by y by z.

PROBLEM. The differential volume element in spherical polar coordinates is $dV = r^2 \sin\theta\, d\phi\, d\theta\, dr$. Given that ϕ goes from 0 to 2π, θ goes from 0 to π, and r goes from 0 to r, evaluate the triple integral

$$V = \int_0^r \int_0^\pi \int_0^{2\pi} r^2 \sin\theta\, d\phi\, d\theta\, dr$$

Solution.

$$\int_0^{2\pi} r^2 \sin\theta\, d\phi = 2\pi r^2 \sin\theta$$

$$\int_0^{\pi} 2\pi r^2 \sin\theta\, d\theta = 4\pi r^2$$

$$\int_0^{r} 4\pi r^2\, dr = \frac{4}{3}\pi r^3 = V$$

SUGGESTED READING

1. BRADLEY, GERALD L., and SMITH, KARL J., *Calculus*, Prentice-Hall, Inc., Upper Saddle River, NJ, 1995.
2. VARBERG, DALE, and PURCELL, EDWIN J., *Calculus*, 7th ed., Prentice-Hall, Inc., Upper Saddle River, NJ, 1997.

PROBLEMS

1. Evaluate the following integrals (consider all uppercase letters to be constants):

(a) $\int 4x^2\, dx$

(b) $\int \frac{1}{x^2}\, dx$

(c) $\int \sin 3x\, dx$

(d) $\int (3x + 5)^2 x\, dx$

(e) $\int 4e^x\, dx$

(f) $\int P\, dv$

(g) $\int \frac{RT}{p}\, dp$

(h) $\int Mv\, dv$

(i) $\int \frac{Q^2}{r^2}\, dr$

(j) $\int \cos\,(2\pi Wt)\, dt$

2. Evaluate the following integrals using the Table of Integrals found in Appendix II, as needed (consider all uppercase letters to be constants):

(a) $\int e^{-4x}\, dx$

(b) $\int (x^2 - A^2)\, dx$

(c) $\int (x^2 - A^2)^{1/2}\, dx$

(d) $\int (x^4 - 2x^2 + 4)\, x^3\, dx$

(e) $\int \sin^2\left(\frac{N\pi x}{A}\right) dx$

(f) $\int \sin^2\left(\frac{N\pi x}{A}\right) x\, dx$

(g) $\int \left(\frac{-\Delta H}{Rt^2}\right) dt$

(h) $\int e^x \cos x \, dx$

(i) $\int \sin^2 (2\pi Wt) \, dt$

(j) $\int \cos^3 \phi \sin \phi \, d\phi$

(k) $\int \cos^4 \theta \, d\theta$

(l) $\int \sin^6 (3x + 4) \, dx$

(m) $\int x^2 \cos 2x \, dx$

(n) $\int \frac{dx}{(4 - x)(3 - x)}$

(o) $\int \left(\frac{\Delta H}{t^2} + \frac{A}{t} + \frac{B}{2} + \frac{C}{3}t\right) dt$

(p) $\int \frac{C_p}{t} \, dt$

(q) $\int \frac{dx}{(A - x)^n}$

(r) $\int r \, e^{-ar} \, dr$

(s) $\int e^{-\epsilon/KT} \, d\epsilon$

(t) $\int \frac{d(A)}{(A)} = -\int K \, dt$

3. Evaluate the following definite integrals using the Table of Indefinite and Definite Integrals found in Appendix II, as needed:

(a) $\int_{T_1}^{T_2} \left(a + bT + cT^2 + \frac{d}{T}\right) dT$; a, b, c, and d constants

(b) $\int_{P_1}^{P_2} \frac{RT}{P} \, dP$; R and T constants

(c) $\int_0^{2\pi} d\phi$

(d) $\int_{T_1}^{T_2} \frac{\Delta H}{RT^2} \, dT$; ΔH and R constants

(e) $\int_{V_1}^{V_2} \left(\frac{nRT}{V - nb} - \frac{n^2 a}{V^2}\right) dV$; a, b, n, R, and T constants

(f) $\int_0^{\pi/2} \sin^2 \theta \cos \theta \, d\theta$

(g) $\int_0^a x^2 \sin^2 \frac{n\pi x}{a} \, dx$; n, π, and a constants

(h) $\int_0^\infty x^2 e^{-ax^2} \, dx$; a constant

(i) $\int_0^\infty e^{-2r/a_0} r \, dr$; a_0 constant

(j) $\int_0^\infty e^{-mv^2/2kT} v^3 \, dv$; m, k, and T constants

(k) $\int_0^\infty (2J + 1) e^{-a(J^2 + J)} \, dJ$; a constant

4. Consider the ideal gas law equation $P = nRT/V$, where in this case n, R, and T are assumed to be constant. Prepare a graph of P versus V, choosing suitable coordinates, for $n =$ 1 mole, $R = 0.0821\ell \cdot$ atm/mol $\cdot$ K, and $T = 298$ K from a volume of $V = 1.00$ liters to a volume of $V = 10.0$ liters. Consider now the area under the P versus V curve from $V =$ 2.00 liters to $V = 6.00$ liters. Determine the approximate area graphically by breaking up the area into four rectangles of equal width ΔV; compare your answer to that found by analytically integrating the function between these limits of integration.
5. Evaluate the following multiple integrals using the Table of Integrals, as needed:

(a) $\int\int yx^2 \, dx \, dy$

(b) $\int\int (x^2 + y^2) \, dx \, dy$

(c) $\int\int y \ln x \, dx \, dy$

(d) $\int\int\int x^2 \ln y \, e^{2x} \, dx \, dy \, dz$

(e) $\int_0^{\pi/2}\int_0^{2} r \cos\theta \, dr \, d\theta$

(f) $\int_0^{2\pi}\int_0^{\pi}\int_0^{r} r^2 \sin\theta \, dr \, d\theta \, d\phi$

(g) $\int_0^{\infty}\int_0^{\infty}\int_0^{\infty} e^{-\frac{h^2}{8mkT}\left(\frac{n_x^2}{a^2}+\frac{n_y^2}{b^2}+\frac{n_z^2}{c^2}\right)} dn_x \, dn_y \, dn_z$; a, b, c, h, m, k, and T constants

6. The equation of a straight line passing through the origin of a Cartesian coordinate system is $y = mx$, where m is the slope of the line. Show that the area of a triangle made up of this line and the x axis between $x = 0$ and $x = a$ is $A = \frac{1}{2}ay$.
7. The Kirchhoff equation for a chemical reaction relating the variation of ΔH of a reaction with absolute temperature is

$$\left[\frac{\partial(\Delta H)}{\partial T}\right]_P = \Delta C_p$$

where ΔC_p represents the change in the heat capacity at constant pressure for the reaction. Expressing ΔC_p as a power series in T,

$$\Delta C_p = a + bT + cT^2$$

derive an equation for ΔH as a function of temperature. (*Hint:* Write the above derivative in differential form.)
8. The Gibbs-Helmholtz equation for a chemical reaction is

$$\left[\frac{\partial(\Delta G/T)}{\partial T}\right]_P = -\frac{\Delta H}{T^2}$$

where ΔG is the Gibbs free energy change attending the reaction, ΔH is the enthalpy change attending the reaction, and T is absolute temperature. Expressing ΔH in a power series in T,

$$\Delta H = a + bT + cT^2$$

where a, b, and c are experimentally determined constants, derive an expression for ΔG as a function of temperature.

9. Find the probability of finding a particle confined to a field-free one-dimensional box in the state $n = 1$ at $x = L/2$ in a range $L/2 \pm 0.05\,L$, where L is the width of box, given

$$\text{Probability} = \frac{2}{L} \int_{L/2-0.05L}^{L/2+0.05L} \sin^2 \frac{\pi x}{L}\, dx$$

10. Find the probability of finding an electron in the 1s-state of the hydrogen atom at $r = a_0$ in a range $a_0 \pm 0.005\,a_0$, where a_0 is the Bohr radius, given

$$\text{Probability} = 4\left(\frac{1}{a_0}\right)^3 \int_{a_0-0.005a_0}^{a_0+0.005a_0} e^{-2r/a_0} r^2\, dr$$

11. Find the expectation value $\langle x \rangle$ for an electron in the 1s-state of the hydrogen atom, given that

$$\langle x \rangle = 4\left(\frac{1}{a_0}\right)^3 \int_0^\infty e^{-2r/a_0} r^3\, dr$$

12. The differential volume element in cylindrical coordinates is $dV = r\, d\theta\, dr\, dz$. Show that if r goes from 0 to r, θ from 0 to 2π, and z from 0 to h, the volume of a cylinder is $V = \pi r^2 h$.

$$\hat{\mathbf{H}}\Psi = E\Psi$$

6

Differential Equations

6-1 INTRODUCTION

The equation

$$f\left(x, y, \frac{dy}{dx}, \frac{d^2y}{dx^2}, \frac{d^3y}{dx^3}, \ldots, \frac{d^ny}{dx^n}\right) = 0 \qquad (6\text{-}1)$$

where $y = f(x)$ is known as a *differential equation*. The *order* of the differential equation is the order of the highest derivative that appears in the equation. Hence,

$$\frac{d^3y}{dx^3} + y^6 = 0$$

is an example of a third-order differential equation.

A linear differential equation is one having the form

$$A_0(x)\frac{d^ny}{dx^n} + A_1(x)\frac{d^{n-1}y}{dx^{n-1}} + \cdots + A_{n-1}(x)\frac{dy}{dx} + A_n(x)y + B(x) = 0 \qquad (6\text{-}2)$$

where $A_0(x) \neq 0$. A third-order linear differential equation, therefore, would have the form

$$A_0(x)\frac{d^3y}{dx^3} + A_1(x)\frac{d^2y}{dx^2} + A_2(x)\frac{dy}{dx} + A_3(x)y + B(x) = 0$$

It is customary to divide through the equation by $A_0(x)$, giving the equation

$$\frac{A_0(x)}{A_0(x)}\frac{d^3y}{dx^3} + \frac{A_1(x)}{A_0(x)}\frac{d^2y}{dx^2} + \frac{A_2(x)}{A_0(x)}\frac{dy}{dx} + \frac{A_3(x)}{A_0(x)}y + \frac{B(x)}{A_0(x)} = 0 \tag{6-3}$$

or

$$\frac{d^3y}{dx^3} + N(x)\frac{d^2y}{dx^2} + P(x)\frac{dy}{dx} + Q(x)y + R(x) = 0 \tag{6-4}$$

When the variable $R(x)$ equals zero, Equation (6-4) is known as the *reduced equation.*

$$\frac{d^3y}{dx^3} + N(x)\frac{d^2y}{dx^2} + P(x)\frac{dy}{dx} + Q(x)y = 0 \tag{6-5}$$

If a function and its derivatives are substituted into Equation (6-1) and satisfy the equation, then the function is said to be a solution to the differential equation.

6-2 LINEAR COMBINATIONS

Suppose that $u(x)$ and $v(x)$ are two solutions to Equation (6-5). We easily can show, then, that a linear combination of the two solutions

$$\phi = c_1u(x) + c_2v(x)$$

where c_1 and c_2 are arbitrary constants, also is a solution to Equation (6-5). If we substitute ϕ and its derivatives into Equation (6-5), we obtain the equation

$$c_1\frac{d^3u}{dx^3} + c_2\frac{d^3v}{dx^3} + c_1N(x)\frac{d^2u}{dx^2} + c_2N(x)\frac{d^2v}{dx^2}$$
$$+ c_1P(x)\frac{du}{dx} + c_2P(x)\frac{dv}{dx} + c_1Q(x)u + c_2Q(x)v = 0$$

Collecting terms gives

$$\begin{aligned} &c_1\left(\frac{d^3u}{dx^3} + N(x)\frac{d^2u}{dx^2} + P(x)\frac{du}{dx} + Q(x)u\right) \\ &\quad + c_2\left(\frac{d^3v}{dx^3} + N(x)\frac{d^2v}{dx^2} + P(x)\frac{dv}{dx} + Q(x)v\right) = 0 \end{aligned} \tag{6-6}$$

However, since both $u(x)$ and $v(x)$ are solutions to Equation (6-5), each term in parentheses in Equation (6-6) is identically equal to zero and the equation is identically satisfied. Thus, ϕ also is a solution to the equation.

In general, then, if $u_1(x)$, $u_2(x)$, $u_3(x)$, . . . are solutions to a linear differential equation, then a linear combination of these solutions, $\phi = c_1u_1 + c_2u_2 + c_3u_3 + \cdots$, also is a solution.

6-3 FIRST-ORDER DIFFERENTIAL EQUATIONS

A first-order linear differential equation is one having the general form

$$\frac{dy}{dx} + Q(x)y + R(x) = 0 \tag{6-7}$$

The reduced form of this equation can be solved by simple integration, using a technique known as *separation of variables*. We put all the y variables on one side of the equation and all the x variables on the other side:

$$\frac{dy}{dx} + Q(x)y = 0$$
$$\frac{dy}{y} = -Q(x)\,dx$$
$$\int \frac{dy}{y} = \int -Q(x)\,dx$$
$$\ln y = -\int Q(x)\,dx + C$$

Taking the antilogarithm of this equation gives

$$y = A\,e^{-\int Q(x)\,dx} \tag{6-8}$$

where $A = e^C$ is a constant.

Examples

1. The rate of a certain chemical reaction is found to be proportional to the concentration of reactant at any time t. Find the integrated rate equation describing such a process.

   Solution. The rate of the reaction can be described by the derivative $-d(A)/dt$, which gives the rate of decrease of the concentration of reactant A. Therefore,

$$-\frac{d(A)}{dt} = k(A)$$

where k is a constant of proportionality. Separating variables gives

$$\int \frac{d(A)}{(A)} = -\int k\,dt = -k \int dt$$

$$\ln(A) = -kt + C$$

We can evaluate the constant of integration by assuming that at $t = 0$, $(A) = (A)_0$, some initial concentration of A, which gives $C = \ln\,(A)_0$. Therefore, we can write

$$\ln\,(A) = -kt + \ln\,(A)_0$$

or

$$(A) = (A)_0\, e^{-kt}$$

which describes the exponential decay typical of a first-order process.

2. Any phase change of a substance taking place at constant pressure and temperature can be described by the Clapeyron equation

$$\frac{dP}{dT} = \frac{\Delta H}{T(V_f - V_i)}$$

where ΔH is the enthalpy change attending the phase change, and V_i and V_f are the molar volumes of the initial and final phases, respectively. Find the integrated form of this equation for the vaporization of a liquid, assuming the vapor to be a perfect gas.

Solution. For a liquid-vapor phase change, the Clapeyron equation becomes

$$\frac{dP}{dT} = \frac{\Delta H_{\text{vap}}}{T(V_g - V_l)}$$

where P is the vapor pressure of the liquid at any temperature T. We assume that V_g is much greater than V_l, which allows us to drop V_l from the equation, and since the vapor is assumed to be a perfect gas, then $V_g = RT/P$. Substituting this back into the Clapeyron equation gives

$$\frac{dP}{dT} = \frac{P\Delta H_{\text{vap}}}{RT^2}$$

Separating variables and assuming ΔH_{vap} is constant, we have

$$\int \frac{dP}{P} = \frac{\Delta H_{\text{vap}}}{R} \int \frac{dT}{T^2}$$

which integrates to give the Clausius-Clapeyron equation

$$\ln P = -\frac{\Delta H_{\text{vap}}}{RT} + C$$

Consider, now, the nonreduced equation written in the form

$$\frac{dy}{dx} + Q(x)y = f(x) \tag{6-9}$$

Equation (6-8) can be used to help us solve Equation (6-9). We observe that

$$\frac{d}{dx}\left[e^{\int Q(x)dx}y\right] = e^{\int Q(x)dx}\frac{dy}{dx} + e^{\int Q(x)dx}Q(x)y \tag{6-10}$$

Therefore, if we multiply Equation (6-9) through by $e^{\int Q(x)dx}$, we have

$$e^{\int Q(x)dx}\frac{dy}{dx} + e^{\int Q(x)dx}Q(x)y = e^{\int Q(x)dx}f(x) \tag{6-11}$$

which now can be integrated. Integration of the left side of Equation (6-11) is found using Equation (6-10). Hence, we have

$$e^{\int Q(x)dx}y = \int e^{\int Q(x)dx}f(x)\,dx + C \tag{6-12}$$

The term $e^{\int Q(x)dx}$ is known as an *integrating factor.*

Example

In the consecutive reaction $A \xrightarrow{k_1} B \xrightarrow{k_2} C$, where k_1 and k_2 are rate constants, the concentration of B, (B), follows the rate equation

$$\frac{d(B)}{dt} = k_1(A)_0e^{-k_1t} - k_2(B)$$

Here $(A)_0$ represents the concentration of A at $t = 0$. Find the integrated rate equation describing the concentration of B as a function of time.

Solution. We first put the rate equation in the form given by Equation (6-9)

$$\frac{d(B)}{dt} + k_2(B) = k_1(A)_0e^{-k_1t}$$

The integrating factor for this equation is $e^{\int k_2dt} = e^{k_2t}$. Multiplying through by the integrating factor gives

$$e^{k_2t}\frac{d(B)}{dt} + k_2e^{k_2t}(B) = k_1(A)_0e^{k_2}e^{-k_1t}$$

The left side of the equation integrates to give $e^{k_2t}(B)$. The right side of the equation is easily integrated if we combine the exponentials.

$$e^{k_2t}(B) = k_1(A)_0\int e^{(k_2-k_1)t}\,dt$$

which gives

$$e^{k_2 t}(B) = \frac{k_1(A)_0}{(k_2 - k_1)} e^{(k_2 - k_1)t} + C$$

Again, the constant of integration can be determined by assuming that at $t = 0$, $(B) = 0$. With this and some algebraic manipulation, we obtain the final equation

$$(B) = \frac{k_1(A)_0}{(k_2 - k_1)} \left(e^{-k_1 t} - e^{-k_2 t} \right)$$

6-4 SECOND-ORDER DIFFERENTIAL EQUATIONS WITH CONSTANT COEFFICIENTS

One type of linear differential equation that is extremely important to the study of physical chemistry, and indeed to the physical sciences as a whole, is the *second-order linear differential equation with constant coefficients*, having the general form

$$\frac{d^2y}{dx^2} + A\frac{dy}{dx} + By = R(x) \tag{6-13}$$

where A and B are constants. Consider, first, the reduced equation, where $R(x) = 0$.

$$\frac{d^2y}{dx^2} + A\frac{dy}{dx} + By = 0 \tag{6-14}$$

A trivial solution to this equation—and, in fact, to all reduced equations, irrespective of order—is $y = 0$. Let us guess a nontrivial solution as[1]

$$y = e^{mx} \neq 0 \tag{6-15}$$

where m is a constant. Taking the first and second derivative of Equation (6-15), we have

$$\frac{dy}{dx} = me^{mx} \quad \text{and} \quad \frac{d^2y}{dx^2} = m^2 e^{mx}$$

Substituting these into Equation (6-14) gives

$$m^2 e^{mx} + Ame^{mx} + Be^{mx} = 0$$

Since $e^{mx} \neq 0$, we can divide through by it, giving

$$m^2 + Am + B = 0 \tag{6-16}$$

[1]One acceptable way to solve a differential equation is to guess a solution (in German called an *Ansatz*) and to substitute the solution and its derivatives into the differential equation to see if it satisfies the equation.

Equation (6-16) is called the *auxiliary equation*. Hence, $y = e^{mx}$ is a solution to Equation (6-14), provided there is an m such that

$$m = \frac{-A \pm \sqrt{A^2 - 4B}}{2} \tag{6-17}$$

Depending on the sign and/or the magnitude of constants A and B, the roots to Equation (6-16) may be real, imaginary, or complex. Let us consider the solutions to the differential equation in each of these cases.

Real Roots. Consider, first, the case in which the roots to the auxiliary equation are real, say $m = \pm a$. Thus,

$$y = e^{ax} \quad \text{and} \quad y = e^{-ax}$$

are each solutions to Equation (6-14), called *particular solutions*. A general solution to Equation (6-14), when the roots to the auxiliary equation are real, is found by taking a linear combination of the two particular solutions.

$$y = c_1 e^{ax} + c_2 e^{-ax} \tag{6-18}$$

where c_1 and c_2 are arbitrary constants. We see, then, that when the roots to the auxiliary equation are real, the solution to the differential equation is real and is the combination of an exponential increase plus an exponential decay.

Imaginary Roots. Consider, next, the case in which the roots to Equation (6-17) are pure imaginary, say, $m = \pm ib$, where $i = \sqrt{-1}$ (see Chapter 1). Thus,

$$y_1 = e^{ibx} \quad \text{and} \quad y_2 = e^{-ibx}$$

are two particular solutions to Equation (6-14). Again, a general solution is found by taking a linear combination of the particular solutions

$$y = c_1 e^{ibx} + c_2 e^{-ibx} \tag{6-19}$$

Recall from Chapter 1 that the exponentials $e^{i\theta}$ and $e^{-i\theta}$ can be related to sine and cosine functions. Without proof, then, another form of the general solution to Equation (6-14), in the case where the roots to the auxiliary equation are imaginary, is

$$y = c_1' \sin bx + c_2' \cos bx \tag{6-20}$$

Since, generally, sine and cosine functions describe waves, it is not surprising that solutions to wave equations have the form given by Equations (6-19) and (6-20).

Complex Roots. Consider, now, the case in which the roots are complex, say, $m = a \pm ib$. In this case we have the two particular solutions

$$y_1 = e^{(a+ib)x} \quad \text{and} \quad y_2 = e^{(a-ib)x}$$

A general solution is, therefore,

$$\begin{aligned} y &= c_1 e^{(a+ib)x} + c_2 e^{(a-ib)x} \\ &= c_1 e^{ax} e^{ibx} + c_2 e^{ax} e^{-ibx} \\ &= e^{ax}(c_1 e^{ibx} + c_2 e^{-ibx}) \end{aligned}$$

Double Roots. There is a possibility that the two roots to the auxiliary equation may be real and equal. In this case, the general solution is

$$y = c_1 e^{ax} + c_2 x e^{ax} \tag{6-21}$$

Examples

1. Solve $\frac{d^2y}{dx^2} - 4\frac{dy}{dx} + 4y = 0$.

To find the auxiliary equation, substitute $y = e^{mx}$ and its derivatives into the equation. This gives

$$\begin{aligned} m^2 - 4m + 4 &= 0 \\ (m-2)^2 &= 0 \end{aligned}$$

which has the real, double roots $m = 2, 2$. The general solution to the equation is

$$y = c_1 e^{2x} + c_2 x e^{2x}$$

2. Solve $\frac{d^2y}{dx^2} + 5y = 0$.

Substituting $y = e^{mx}$ and its derivatives into this equation gives

$$m^2 + 5 = 0$$

which has the imaginary roots $m = \pm i\sqrt{5}$. The general solution to the equation is

$$y = c_1 e^{i\sqrt{5}x} + c_2 e^{-i\sqrt{5}x}$$

or in the sine-cosine form

$$y = c_1' \sin\sqrt{5}x + c_2' \cos\sqrt{5}x$$

3. Solve $\frac{d^2\psi}{dx^2} + \frac{8\pi^2 mE}{h^2}\psi = 0$, where π, m, h, and E are constants.

Let $\psi = e^{nx}$. Substituting this and its derivatives into the equation gives the auxiliary equation

$$n^2 + \frac{8\pi^2 mE}{h^2} = 0$$

which has the imaginary roots $n = +i\sqrt{\frac{8\pi^2 mE}{h^2}}$ and $n = -i\sqrt{\frac{8\pi^2 mE}{h^2}}$.
Choosing the sine-cosine form of the solution, we have

$$\psi = A\sin\sqrt{\frac{8\pi^2 mE}{h^2}}x + B\cos\sqrt{\frac{8\pi^2 mE}{h^2}}x$$

where A and B are constants.

6-5 GENERAL SERIES METHOD OF SOLUTION

In certain cases, differential equations cannot be solved by the simple method outlined in Section 6-4. For this reason, consider another important method for solving differential equations, called the *series method* of solution. Generally, this method is used for equations that are not reduced and for equations where the coefficients are not constant. To see how it works, let us introduce the method on a very simple reduced equation.

Consider the equation

$$\frac{d^2y}{dx^2} + \beta^2 y = 0 \tag{6-22}$$

where β is a constant. We easily can show, using the method outlined in Section 6-4, that two particular solutions of Equation (6-22) are $\sin \beta x$ and $\cos \beta x$. Let us assume, however, that the solution to Equation (6-22) is a series of the form

$$y = a_\kappa x^\kappa + a_{\kappa+1}x^{(\kappa+1)} + a_{\kappa+2}x^{(\kappa+2)} + \cdots = \sum_{n=0}^{\infty} a_{\kappa+n}x^{(\kappa+n)} \tag{6-23}$$

where κ represents the lowest power that x can have in the summation. Taking the first and second derivatives of this equation gives

$$\frac{dy}{dx} = \sum_{n=1}^{\infty}(\kappa+n)a_{\kappa+n}x^{(\kappa+n-1)} \quad \text{and}$$

$$\frac{d^2y}{dx^2} = \sum_{n=2}^{\infty}(\kappa+n-1)(\kappa+n)a_{\kappa+n}x^{(\kappa+n-2)}$$

Substituting the second derivative and y back into Equation (6-22), we have

$$\sum_{n=2}^{\infty}(\kappa+n-1)(\kappa+n)a_{\kappa+n}x^{(\kappa+n-2)} + \beta^2\sum_{n=0}^{\infty}a_{\kappa+n}x^{(\kappa+n)} = 0 \tag{6-24}$$

Equation (6-24) must hold for every value of x, and this will be true only if every coefficient of the power of x is identically equal to zero. Since n cannot be negative, the lowest power of x in the first summation in Equation (6-24) is $x^{(\kappa-2)}$, where $n = 0$. Substituting this into the first summation gives

$$(\kappa - 1)\kappa a_\kappa = 0 \tag{6-25}$$

Equation (6-25) is called an *indicial equation*. Since $a_\kappa \neq 0$, $(\kappa - 1)\kappa = 0$, which gives

$$\kappa = 0, 1$$

We can combine the summations in Equation (6-24) if we replace n in the first summation with $n + 2$, giving[2]

$$\sum_{n=0}^{\infty} (\kappa + n + 1)(\kappa + n + 2) a_{\kappa+n+2} x^{(\kappa+n)} + \beta^2 \sum_{n=0}^{\infty} a_{\kappa+n} x^{(\kappa+n)} = 0$$

$$\sum_{n=0}^{\infty} \left\{ (\kappa + n + 1)(\kappa + n + 2) a_{\kappa+n+2} + \beta^2 a_{\kappa+n} \right\} x^{(\kappa+n)} = 0$$

Since $x^{(\kappa+n)} \neq 0$, the term in { } must equal zero, giving

$$a_{\kappa+n+2} = \frac{-\beta^2 a_{\kappa+n}}{(\kappa + n + 1)(\kappa + n + 2)} \tag{6-26}$$

Equation (6-26) is called a *recursion equation* or *recursion formula*; it connects the coefficients of the series. We see in this case that we have two series expansions: one for which $\kappa = 0$ and one for which $\kappa = 1$. Let us look at the $\kappa = 0$ expansion. Under these circumstances,

$$a_{n+2} = \frac{-\beta^2 a_n}{(n + 1)(n + 2)}$$

$$a_2 = \frac{-\beta^2 a_0}{2}, \quad a_4 = \frac{\beta^4 a_0}{24}, \ldots$$

With $a_0 = 1$, we have

$$y = 1 - \frac{\beta^2}{2} x^2 + \frac{\beta^4}{24} x^4 - + \cdots = 1 - \frac{(\beta x)^2}{2} + \frac{(\beta x)^4}{24} - + \cdots$$

which we recognize as the series expansion for cos βx, a particular solution of Equation (6-22). Substituting $\kappa = 1$ into Equation (6-26), and letting $a_1 = \beta$, it is easy to show that the series which results is the series expansion for sin βx, another particular solution to Equation (6-22).

[2]Adding two more terms to an infinite sum is the same as adding zero.

6-6 SPECIAL POLYNOMIAL SOLUTIONS TO DIFFERENTIAL EQUATIONS

Hermite's Equation. Consider the differential equation

$$\frac{d^2y}{dx^2} - 2x\frac{dy}{dx} + 2\alpha y = 0 \tag{6-27}$$

Equation (6-27) is known as Hermite's equation. The roots of the indicial equation in this case are $\kappa = 0, 1$ (see Section 6-5), which gives rise to the recursion formula

$$a_{n+2} = \frac{2(\kappa + n) - 2\alpha}{(\kappa + n + 2)(\kappa + n + 1)} a_n$$

For $\kappa = 0$, we have

$$y = a_0 \left(1 - \frac{2\alpha}{2!}x^2 + \frac{2^2\alpha(\alpha - 2)}{4!}x^4 - \frac{2^3\alpha(\alpha - 2)(\alpha - 4)}{6!}x^6 + \cdots\right)$$

For $\kappa = 1$, we have

$$y = a_0 x \left(1 - \frac{2(\alpha - 1)}{3!}x^2 + \frac{2^2(\alpha - 1)(\alpha - 3)}{5!}x^4 - \cdots\right)$$

The general solution to Equation (6-27) is the superposition of these two particular solutions. The appropriate choice of a_0

$$a_0 = (-1)^{n/2} \frac{n!}{\left(\frac{n}{2}\right)!}$$

leads to a set of solutions known as the *Hermite polynomials* of degree n.

$$H_n(x) = (2x)^n - \frac{n(n-1)}{1!}(2x)^{n-2} + \frac{n(n-1)(n-2)(n-3)}{2!}(2x)^{n-4} - \cdots \tag{6-28}$$

Example

The Schrödinger equation describing a simple one-dimensional harmonic oscillator is

$$\frac{d^2\psi}{dx^2} + \frac{8\pi^2 m}{h^2}\left(E - \frac{1}{2}kx^2\right)\psi = 0 \tag{6-29}$$

where π, m, h, E, and k are constants. Show that this equation can be solved using Hermite polynomials.

Solution. Let

$$\varepsilon = \frac{8\pi^2 mE}{h^2} \quad \text{and} \quad \beta^2 = \frac{4\pi^2 mk}{h^2}$$

Substituting these into Equation (6-29) gives

$$\frac{d^2\psi}{dx^2} + (\varepsilon - \beta^2 x^2)\psi = 0$$

We now make a change of variables by letting $\xi = \sqrt{\beta}x$.

$$\frac{d^2\psi}{dx^2} = \left(\frac{\partial^2\psi}{\partial\xi^2}\right)\left(\frac{\partial^2\xi}{\partial x^2}\right) = \sqrt{\beta}\left(\frac{\partial^2\psi}{\partial\xi^2}\right)$$

This gives

$$\frac{d^2\psi}{d\xi^2} + \left[1 - \xi^2 + \left(\frac{\varepsilon}{\beta} - 1\right)\right]\psi = 0 \tag{6-30}$$

We now let $\psi(\xi) = e^{-\xi^2/2}y(\xi)$. Taking the second derivative of this function [see Chapter 4, Section 4-2, item 11, Example (d)], and substituting it and ψ into Equation (6-30) gives

$$\frac{d^2y}{d\xi^2} - 2x\frac{dy}{dx} + \left(\frac{\varepsilon}{\beta} - 1\right)y = 0$$

which is just Equation (6-27), Hermite's equation, the solutions of which are the Hermite polynomials.

Laguerre's Equation. Consider the differential equation

$$x\frac{d^2y}{dx^2} + (1 - x)\frac{dy}{dx} + \alpha y = 0 \tag{6-31}$$

where α is a constant. This equation is known as *Laguerre's equation*. The indicial equation for Laguerre's equation has a single root, $\kappa = 0$, which gives rise to the recursion formula

$$a_{n+1} = \frac{n - \alpha}{(n+1)^2}a_n \tag{6-32}$$

The series solution is therefore

$$y = a_0\left(1 - \alpha x + \frac{\alpha(\alpha - 1)}{(2!)^2}x^2 - \cdots\right) \tag{6-33}$$

Again, suitable choice of a_0,

$$a_0 = (-1)^n n!$$

leads to the *Laguerre polynomials* of degree n:

$$L_n(x) = (-1)^n \left(x^n - \frac{n^2}{1!}x^{n-1} + \frac{n^2(n-1)^2}{2!}x^{n-2} + \cdots + (-1)^n n! \right) \qquad (6\text{-}34)$$

A differential equation closely related to Laguerre's equation is the equation

$$x\frac{d^2y}{dx^2} + (k+1-x)\frac{dy}{dx} + (\alpha - k)y = 0 \qquad (6\text{-}35)$$

where k is an integer ≥ 0. This equation is produced when Equation (6-31) is differentiated k times and y is replaced by the kth derivative. Solutions to this equation are usually represented as

$$y = \frac{d^k}{dx^k}L_n(x) \equiv L_n^k(x)$$

and are called the *associated Laguerre polynomials* of degree $(n-k)$.

A third form of the Laguerre equation, important in the wave-mechanical solution of the radial part of the hydrogen atom, is the equation

$$x\frac{d^2y}{dx^2} + 2\frac{dy}{dx} + \left[n - \frac{k-1}{2} - \frac{x}{4} - \frac{k^2-1}{4x} \right] y = 0 \qquad (6\text{-}36)$$

This equation can be transformed into one having the same form as Equation (6-35) by letting $y = e^{-x/2}x^{(k-1)/2}v$. Substituting y and its derivatives into Equation (6-36) yields the equation

$$x\frac{d^2v}{dx^2} + (k+1-x)\frac{dv}{dx} + (n-k)v = 0 \qquad (6\text{-}37)$$

which we see has exactly the same form as Equation (6-35). Thus, $v = L_n^k$, and a particular solution to Equation (6-36) is

$$y = e^{-x/2}x^{(k-1)/2}L_n^k \qquad (6\text{-}38)$$

This function is called the *associated Laguerre function*.

Example

The radial part of the Schrödinger equation for the hydrogen atom is

$$\frac{1}{r^2}\frac{d}{dr}\left(r^2\frac{dR}{dr} \right) + \left[\frac{2\mu}{\hbar^2}\left(E + \frac{e^2}{4\pi\varepsilon_0 r} \right) - \frac{\ell(\ell+1)}{r^2} \right] R(r) = 0$$

where π, μ, $\hbar$, ε_0, and ℓ are constants. Show that solutions to this equation are the associated Laguerre polynomials.

Solution. To transform the radial equation into a form that resembles Laguerre's equation, let us first expand the equation.

$$\frac{1}{r^2}\frac{d}{dr}\left(r^2\frac{dR}{dr}\right)+\left[\frac{2\mu E}{\hbar^2}+\frac{2\mu e^2}{4\pi\varepsilon_o r}-\frac{\ell(\ell+1)}{r^2}\right]R(r)=0$$

Next, we define two new constants

$$\alpha^2=-\frac{2\mu E}{\hbar^2}\quad\text{and}\quad\beta=\frac{\mu e^2}{4\pi\epsilon_o\alpha}$$

Substituting these into the equation, we have

$$\frac{1}{r^2}\frac{d}{dr}\left(r^2\frac{dR}{dr}\right)+\left[-\alpha^2+\frac{2\beta\alpha}{r}-\frac{\ell(\ell+1)}{r^2}\right]R(r)=0 \qquad \text{(6-39)}$$

We next make a transformation of variables, $\rho = 2\alpha r$.

$$\frac{dR}{dr}=\frac{d\rho}{dr}.\frac{dR}{d\rho}=2\alpha\frac{dR}{d\rho}.\quad\text{Likewise,}\quad\frac{d}{dr}=2\alpha\frac{d}{d\rho}$$

Substituting these back into Equation (6-39) gives

$$\left(\frac{4\alpha^2}{\rho^2}\right)2\alpha\frac{d}{d\rho}\left(\frac{\rho^2}{4\alpha^2}(2\alpha)\frac{dR}{d\rho}\right)+\left(-\alpha^2+\frac{4\beta\alpha^2}{\rho}-(4\alpha^2)\frac{\ell(\ell+1)}{\rho^2}\right)R(\rho)=0$$

Dividing through the equation by $4\alpha^2$,

$$\left(\frac{1}{\rho^2}\right)\frac{d}{d\rho}\left(\rho^2\frac{dR}{d\rho}\right)+\left[-\frac{1}{4}+\frac{\beta}{\rho}-\frac{\ell(\ell+1)}{\rho^2}\right]R(\rho)=0$$

Next, expanding the derivative $d/d\rho$, and remembering that the derivative must be differentiated as a product, we have

$$\frac{1}{\rho}\left[\rho\frac{d^2R}{d\rho^2}+2\frac{dR}{d\rho}\right]+\left[-\frac{1}{4}+\frac{\beta}{\rho}-\frac{\ell(\ell+1)}{\rho^2}\right]R(\rho)=0$$

Finally, multiplying through by ρ,

$$\rho\frac{d^2R}{d\rho^2}+2\frac{dR}{d\rho}+\left[-\frac{\rho}{4}+\beta-\frac{\ell(\ell+1)}{\rho}\right]R(\rho)=0 \qquad \text{(6-40)}$$

which we see has exactly the same form as Equation (6-36). The solutions to Equation (6-40), then, are

$$R(\rho)=\rho^\ell e^{-\rho/2}L^k(\rho),\ k=2\ell+1$$

where $L^k(\rho)$ are the associated Laguerre polynomials.

Legendre's Equation. Consider, now, an equation having the general form

$$(1-x^2)\frac{d^2y}{dx^2}-2x\frac{dy}{dx}+\ell(\ell+1)y=0 \qquad \text{(6-41)}$$

where ℓ is a constant. This equation is known as *Legendre's equation.* Series solution of this equation leads to the indicial equation having roots $\kappa = 0$, 1, and the recursion formula

$$a_{n+2} = \frac{(\kappa + n)(\kappa + n + 1) - \ell(\ell + 1)}{(\kappa + n + 1)(\kappa + n + 2)} a_n \tag{6-42}$$

Like Hermite's equation, we have a choice of even or odd solutions. When $\kappa = 0$, the significant series solution to Legendre's equation is

$$y = \left[1 - \frac{\ell(\ell + 1)}{2!} x^2 + \frac{\ell(\ell - 1)(\ell + 1)(\ell + 3)}{4!} x^4 + \cdots\right] a_0 \tag{6-43}$$

When $\kappa = 1$, the significant series solution is

$$y = \left[x - \frac{(\ell - 1)(\ell + 2)}{3!} x^3 + \frac{(\ell - 1)(\ell - 3)(\ell + 2)(\ell + 4)}{5!} x^5 + \cdots\right] a_1 \tag{6-44}$$

As in the previous special equations, series solutions to differential equations are of particular interest when the series converges to a polynomial. In the case of Legendre's equation the "nuts and bolts" of finding the conditions for convergence are involved and beyond the scope of this text. Students interested in the procedure are referred to the readings listed at the end of the chapter. It is sufficient to say that the general series solution, which is a linear combination of Equations (6-43) and (6-44), reduces to a polynomial when ℓ is an even or odd, positive or negative integer, including zero. Under these conditions, the resulting polynomials P_ℓ, called the *Legendre polynomials*, have the form

$$\begin{aligned} P_\ell(x) = & \frac{1 \cdot 3 \cdot 5 \cdots (2\ell - 1)}{\ell!} \left\{ x^\ell - \frac{\ell(\ell - 1)}{2(2\ell - 1)} x^{\ell - 2} \right. \\ & \left. + \frac{\ell(\ell - 1)(\ell - 2)(\ell - 3)}{2 \cdot 4(2\ell - 1)(2\ell - 3)} x^{\ell - 4} - \cdots \right\} \end{aligned} \tag{6-45}$$

An equation closely related to Legendre's equation, and important in the solution to problems involving rotational motion, is

$$(1 - x^2)\frac{d^2y}{dx^2} - 2x\frac{dy}{dx} + \left(\ell(\ell + 1) - \frac{m^2}{1 - x^2}\right) y = 0 \tag{6-46}$$

where ℓ and m are integers. This equation is known as the *associated Legendre's equation* and has the particular solution

$$y = (1 - x^2)^{m/2} \frac{d^m}{dx^m} P_\ell(x) \tag{6-47}$$

This solution is known as the *associated Legendre function*, or the *associated spherical harmonics*, because it is related to the allowed standing waves on the surface of a sphere.

Example

The $\Theta(\theta)$ equation, an angular part of the Schrödinger equation describing the hydrogen atom, can be expressed as

$$\frac{1}{\sin\theta}\frac{d}{d\theta}\left(\sin\theta\frac{d\Theta}{d\theta}\right)+\left[\ell(\ell+1)-\frac{m^2}{\sin^2\theta}\right]\Theta(\theta)=0$$

where ℓ and m are integers. Show that solutions to this equation are the associated Legendre polynomials.

Solution. We can put the above equation in the form of the associated Legendre's equation if we let $x = \cos\theta$. Therefore, $\sin^2\theta = 1 - \cos^2\theta = 1 - x^2$. Also,

$$\frac{d\Theta}{d\theta}=\frac{dx}{d\theta}\cdot\frac{d\Theta}{dx}=-\sin\theta\frac{d\Theta}{dx} \quad \text{and} \quad \frac{d}{d\theta}=-\sin\theta\frac{d}{dx}$$

Substituting these into the θ equation gives

$$\frac{d}{dx}\left((1-x^2)\frac{d\Theta}{dx}\right)+\left[\ell(\ell+1)-\frac{m^2}{1-x^2}\right]\Theta(x)=0$$

Expanding the derivative d/dx, remembering that $(1 - x^2)\, d\Theta/dx$ must be differentiated as a product, we have

$$(1-x^2)\frac{d^2\Theta}{dx^2}-2x\frac{d\Theta}{dx}+\left(\ell(\ell+1)-\frac{m^2}{1-x^2}\right)\Theta(x)=0$$

which is just the associated Legendre's equation.

6-7 EXACT AND INEXACT DIFFERENTIALS

The expression $M(x, y)\, dx + N(x, y)\, dy$ is said to be an *exact differential* if there exists a function $F(x, y)$ for which

$$dF=\left(\frac{\partial F}{\partial x}\right)_y dx+\left(\frac{\partial F}{\partial y}\right)_x dy=M(x,y)\,dx+N(x,y)\,dy \qquad (6\text{-}48)$$

If $F(x, y)$ does not exist, $M(x, y)\, dx + N(x, y)\, dy$ is not exact and is called an *inexact differential*.

Euler's Test for Exactness. If $M(x, y)\, dx + N(x, y)\, dy$ is exact, then

$$\left(\frac{\partial F}{\partial x}\right)_y=M(x,y) \quad \text{and} \quad \left(\frac{\partial F}{\partial y}\right)_x=N(x,y)$$

Taking the mixed second derivatives gives

$$\left(\frac{\partial^2 F}{\partial y \partial x}\right) = \left(\frac{\partial M}{\partial y}\right)_x \quad \text{and} \quad \left(\frac{\partial^2 F}{\partial x \partial y}\right) = \left(\frac{\partial N}{\partial x}\right)_y$$

However, since

$$\frac{\partial^2 F}{\partial y\, \partial x} = \frac{\partial^2 F}{\partial x\, \partial y}$$

then

$$\left(\frac{\partial M}{\partial y}\right)_x = \left(\frac{\partial N}{\partial x}\right)_y \tag{6-49}$$

is a necessary condition for exactness. Equation (6-49) is known as Euler's (read "oiler's") test for exactness.

Example

Show that the expression

$$dF = 3x^2y^3\, dx + 3y^2x^3\, dy$$

is exact.

Solution. Since $M(x, y) = 3x^2y^3$ and $N(x, y) = 3y^2x^3$, then $\partial M/\partial y = 9x^2y^2$ and $\partial N/\partial x = 9x^2y^2$. Thus, $\partial M/\partial y = \partial N/\partial x$, Euler's test is satisfied, and the differential is exact.

The equation $(\partial F/\partial x)_y = M(x, y)$ can be written in a more general form

$$F(x, y) = \int M(x, y)\, dx + K(y)$$

where $K(y)$ is independent of the variable x. Taking the partial derivative of $F(x, y)$ with respect to y gives

$$\left(\frac{\partial F}{\partial y}\right)_x = \frac{\partial}{\partial y}\int M(x, y)\, dx + \frac{\partial K}{\partial y} = N(x, y)$$

or

$$\frac{\partial K}{\partial y} = N(x, y) - \frac{\partial}{\partial y}\int M(x, y)\, dx \tag{6-50}$$

Integrating Equation (6-50) gives

$$K(y) = \int \left[N(x, y) - \frac{\partial}{\partial y}\int M(x, y)\, dx\right] dy$$

which gives

$$F(x, y) = \int M(x, y)\, dx + \int \left[N(x, y) - \frac{\partial}{\partial y}\int M(x, y)\, dx\right] dy \tag{6-51}$$

Example

Show that the equation

$$dF = 3x^2y^2\,dx + 2x^3y\,dy$$

is exact and determine $F(x, y)$.

Solution.

$$M(x, y) = 3x^2y^2 \quad \text{and} \quad N(x, y) = 2x^3y$$

$$\frac{\partial M}{\partial y} = 6x^2y \quad \text{and} \quad \frac{\partial N}{\partial x} = 6x^2y$$

$$\frac{\partial M}{\partial y} = \frac{\partial N}{\partial x}$$

Euler's test is satisfied and the differential dF is exact. Now,

$$\int M(x, y)\,dx = \int 3x^2y^2\,dx = x^3y^2$$

and

$$\frac{\partial}{\partial y}\int M(x, y)\,dx = 2x^3y$$

Therefore,

$$K(y) = \int \left[N(x, y) - \frac{\partial}{\partial y}\int M(x, y)\,dx\right] dy$$
$$= \int (2x^3y - 2x^3y)\,dy = C$$

which gives

$$F(x, y) = x^3y^2 + C$$

In Chapter 5 we introduced the concept of the *line integral* as representing the area under a curve (or path) taking some function $f(x)$ from x_1 to x_2. Exact and inexact differentials are directly related to line integrals and occupy a significant place in physical chemistry. If du is an exact differential, then the line integral $_L\int_{u_1}^{u_2} du$, which represents the sum of the infinitesimal elements du taking the function u from u_1 to u_2, depends only on the limits of integration. That is, if du is exact, then

$$_L\int_{u_1}^{u_2} du = u_2 - u_1 = \Delta u$$

When u represents a physical variable describing a system, then that variable is said to be a state function, since the function depends only on the initial and final states of the system and not on the path taking the system from state u_1 to u_2.

If, on the other hand, du is an inexact differential, then ${}_L\int_{u_1}^{u_2} du = u$, which is a quantity that depends on the specific path taking u_1 to u_2. The physical variable represented by u is *not* a function of state, and unless the functional relationship between the variables x and y in du is known, the integral ${}_L\int_{u_1}^{u_2} du$ cannot be evaluated. For example, we saw in Chapter 5 that the work done by the expansion of a gas against an external pressure is a function of both the external pressure on the gas (part of the surroundings) and the volume of the gas. We can write this as a total differential

$$dw = M(P_{\text{ext}}, V)\, dP_{\text{ext}} + N\,(P_{\text{ext}}, V)\, dV = -P_{\text{ext}}\, dV$$

Clearly, dw is not an exact differential, since $M(P_{\text{ext}}, V) = 0$ and $N(P_{\text{ext}}, V) = -P_{\text{ext}}$, and Euler's test for exactness is not satisfied. The integral $\int dw = -\int P_{ext}\, dV$ cannot be evaluated, unless the dependence of the external pressure upon the volume of the gas is known. There is, however, no necessary relationship between the external pressure and the volume of the gas; therefore, an infinite number of paths exist taking the gas from $P_1V_1T_1$ to $P_2V_2T_2$. Each path is associated with a specific amount of work.

Another important property of exact differentials is that the integral of an exact differential around a closed path must be equal to zero. That is, if we integrate an exact differential, such as the differential for energy dE, from some initial energy state E_1 around a closed cycle ending up again at state E_1, then

$${}_L\int_{E_1}^{E_1} dE = \oint dE = E_1 - E_1 = 0 \tag{6-52}$$

Such an integral is called a *cyclic integral*. Equation (6-52) can be thought of as the consummate test for a state function.

6-8 INTEGRATING FACTORS

In certain cases an inexact differential can be made exact by multiplying it by a nonzero function called an *integrating factor*. For example, consider the differential equation

$$du = 2y^4\, dx + 4xy^3\, dy \tag{6-53}$$

Applying Euler's test for exactness, we obtain the equations

$$\frac{\partial M}{\partial y} = 8y^3 \quad \text{and} \quad \frac{\partial N}{\partial x} = 4y^3$$

Clearly, Euler's test for exactness is not satisfied, and the differential du is not exact. If we multiply Equation (6-53) by x, we obtain the equation

$$dF = 2xy^4\, dx + 4x^2y^3\, dy \tag{6-54}$$

Note now that

$$\frac{\partial M}{\partial y} = 8xy^3 \quad \text{and} \quad \frac{\partial N}{\partial x} = 8xy^3$$

and the differential is exact. The factor x in this case is an integrating factor, since it transformed the inexact differential du into the exact differential dF. Moreover, since dF is an exact differential, Equation (6-54) is said to be an exact differential equation and can be solved for $F(x, y)$ by the method outlined in the previous section.

6-9 PARTIAL DIFFERENTIAL EQUATIONS

A partial differential equation is one containing partial derivatives and, therefore, more than one independent variable. An example of a partial differential equation is

$$\frac{\partial^2 u}{\partial x^2} + \frac{\partial^2 u}{\partial y^2} + u(x, y) = 0 \tag{6-55}$$

where $u = f(x, y)$. Since u is a function of two variables x and y, the partial differential equation cannot be solved by direct integration. Before a solution can be found, the variables x and y must be separated; but that may not always be possible. One way to separate variables is to assume that the solution is a product of functions of single variables. Let us attempt this type of solution on Equation (6-55). Let

$$u(x, y) = f(x)g(y) \tag{6-56}$$

Taking the partial derivatives of Equation (6-55), we have

$$\frac{\partial u}{\partial x} = g(y)\frac{\partial f}{\partial x} \quad \text{and} \quad \frac{\partial^2 u}{\partial x^2} = g(y)\frac{\partial^2 f}{\partial x^2}$$
$$\frac{\partial u}{\partial y} = f(x)\frac{\partial g}{\partial y} \quad \text{and} \quad \frac{\partial^2 u}{\partial y^2} = f(x)\frac{\partial^2 g}{\partial y^2}$$

Substituting these into Equation (6-55) gives

$$g(y)\frac{\partial^2 f}{\partial x^2} + f(x)\frac{\partial^2 g}{\partial y^2} + f(x)g(y) = 0$$

Dividing through by $f(x)g(y)$ gives

$$\frac{1}{f(x)}\frac{\partial^2 f}{\partial x^2} + \frac{1}{g(y)}\frac{\partial^2 g}{\partial y^2} + 1 = 0 \tag{6-57}$$

Rearranging Equation (6-57), we have

$$\frac{1}{f(x)}\frac{\partial^2 f}{\partial x^2} + \frac{1}{g(y)}\frac{\partial^2 g}{\partial y^2} = -1 \tag{6-58}$$

Notice that the first term in Equation (6-58) is only a function of the variable x, and the second term in the equation is only a function of the variable y. Hence, the two variables x and y have been separated. Further note that these two terms

$$\frac{1}{f(x)}\frac{\partial^2 f}{\partial x^2} \quad \text{and} \quad \frac{1}{g(y)}\frac{\partial^2 g}{\partial y^2}$$

must each equal a constant for the following reason: If we assume that $f(x)$ is the variable, and differentiate partially with respect to x, then we must hold $g(y)$ and its derivatives constant, but that means that $(1/f(x))(\partial^2 f/\partial x^2)$ also must be constant, since a variable cannot be equal to the sum of two constants. Therefore,

$$\frac{1}{f(x)}\frac{d^2 f}{dx^2} = k_1 \quad \text{and} \quad \frac{1}{g(y)}\frac{d^2 g}{dy^2} = k_2$$

where $k_1 + k_2 = -1$. The equations

$$\frac{1}{f(x)}\frac{d^2 f}{dx^2} - k_1 = 0 \quad \text{and} \quad \frac{1}{g(y)}\frac{d^2 g}{dy^2} - k_2 = 0$$

can be solved by methods outlined in previous sections of this chapter. Once $f(x)$ and $g(y)$ are known, the solution to Equation (6-55) is $u(x, y) = f(x)g(y)$.

Problem. Separate the Schrödinger equation describing the hydrogen atom into three equations: an $R(r)$ equation, a $\Theta(\theta)$ equation, and a $\Phi(\phi)$ equation.

$$\frac{1}{r^2}\frac{\partial}{\partial r}\left(r^2\frac{\partial\psi}{\partial r}\right) + \frac{1}{r^2\sin\theta}\frac{\partial}{\partial\theta}\left(\sin\theta\frac{\partial\psi}{\partial\theta}\right) + \frac{1}{r^2\sin^2\theta}\frac{\partial^2\psi}{\partial\phi^2} + \frac{2\mu}{\hbar^2}\left(E + \frac{e^2}{4\pi\varepsilon_0 r}\right)\psi(r,\theta,\phi) = 0 \tag{6-59}$$

where π, ε_0, $\hbar$, μ, and e are constants.

Solution. Assume a product solution $\psi(r, \theta, \phi) = R(r)\Theta(\theta)\Phi(\phi)$. Evaluating the derivatives, we have

$$\frac{\partial\psi}{\partial r} = \Theta(\theta)\Phi(\phi)\frac{\partial R}{\partial r}, \quad \frac{\partial\psi}{\partial\theta} = R(r)\Phi(\phi)\frac{\partial\Theta}{\partial\theta}, \quad \text{and} \quad \frac{\partial^2\psi}{\partial\phi^2} = R(r)\Theta(\theta)\frac{\partial^2\Phi}{\partial\phi^2}$$

Substituting these into Equation (6-59) gives

$$\frac{\Theta(\theta)\Phi(\phi)}{r^2}\frac{\partial}{\partial r}\left(r^2\frac{\partial R}{\partial r}\right) + \frac{R(r)\Phi(\phi)}{r^2\sin\theta}\frac{\partial}{\partial\theta}\left(\sin\theta\frac{\partial\Theta}{\partial\theta}\right) + \frac{R(r)\Theta(\theta)}{r^2\sin^2\theta}\frac{\partial^2\Phi}{\partial\phi^2} + \frac{2\mu}{\hbar^2}\left(E + \frac{e^2}{4\pi\varepsilon_0 r}\right)R(r)\Theta(\theta)\Phi(\phi) = 0$$

Dividing through by $R(r)\Theta(\theta)\Phi(\phi)$, we have

$$\frac{1}{R(r)}\frac{1}{r^2}\frac{\partial}{\partial r}\left(r^2\frac{\partial R}{\partial r}\right)+\frac{1}{\Theta(\theta)}\frac{1}{r^2\sin\theta}\frac{\partial}{\partial\theta}\left(\sin\theta\frac{\partial\Theta}{\partial\theta}\right)$$
$$+\frac{1}{\Phi(\phi)}\frac{1}{r^2\sin^2\theta}\frac{\partial^2\Phi}{\partial\phi^2}+\frac{2\mu}{\hbar^2}\left(E+\frac{e^2}{4\pi\varepsilon_0 r}\right)=0 \tag{6-60}$$

Next, we multiply through Equation (6-60) by $r^2\sin^2\theta$.

$$\frac{\sin^2\theta}{R(r)}\frac{\partial}{\partial r}\left(r^2\frac{\partial R}{\partial r}\right)+\frac{\sin\theta}{\Theta(\theta)}\frac{\partial}{\partial\theta}\left(\sin\theta\frac{\partial\Theta}{\partial\theta}\right)+\frac{1}{\Phi(\phi)}\frac{\partial^2\Phi}{\partial\phi^2}$$
$$+\frac{2\mu r^2\sin^2\theta}{\hbar^2}\left(E+\frac{e^2}{4\pi\varepsilon_0 r}\right)=0$$

Notice that this isolates the ϕ term. By the same arguments used above, the ϕ term must equal a constant, call it $-m^2$. Therefore,

$$\frac{1}{\Phi(\phi)}\frac{d^2\Phi}{d\phi^2}=-m^2$$

and

$$\frac{\sin^2\theta}{R(r)}\frac{\partial}{\partial r}\left(r^2\frac{\partial R}{\partial r}\right)+\frac{\sin\theta}{\Theta(\theta)}\frac{\partial}{\partial\theta}\left(\sin\theta\frac{\partial\Theta}{\partial\theta}\right)+\frac{2\mu r^2\sin^2\theta}{\hbar^2}\left(E+\frac{e^2}{4\pi\varepsilon_0 r}\right)=m^2$$

since the sum of these two terms equals zero. Dividing through this equation by $\sin^2\theta$ will separate the equation into r terms and θ terms.

$$\frac{1}{R(r)}\frac{\partial}{\partial r}\left(r^2\frac{\partial R}{\partial r}\right)+\frac{1}{\Theta(\theta)}\frac{1}{\sin\theta}\frac{\partial}{\partial\theta}\left(\sin\theta\frac{\partial\Theta}{\partial\theta}\right)$$
$$+\frac{2\mu r^2}{\hbar^2}\left(E+\frac{e^2}{4\pi\varepsilon_0 r}\right)-\frac{m^2}{\sin^2\theta}=0$$

Letting the combined r terms equal a constant $\ell(\ell+1)$, we have

$$\frac{1}{R(r)}\frac{d}{dr}\left(r^2\frac{dR}{dr}\right)+\frac{2\mu r^2}{\hbar^2}\left(E+\frac{e^2}{4\pi\varepsilon_0 r}\right)=\ell(\ell+1)$$

which we recognize as Laguerre's equation. The combined θ terms will equal $-\ell(\ell+1)$

$$\frac{1}{\Theta(\theta)}\frac{1}{\sin\theta}\frac{d}{d\theta}\left(\sin\theta\frac{d\Theta}{d\theta}\right)-\frac{m^2}{\sin^2\theta}=-\ell(\ell+1)$$

which we recognize as the associated Legendre's equation.

SUGGESTED READING

1. BRADLEY, GERALD L., and SMITH, KARL J., *Calculus*, Prentice-Hall, Inc., Upper Saddle River, NJ, 1995.
2. MARGENAU, HENRY, and MURPHY, GEORGE, *The Mathematics of Physics and Chemistry*, D. van Nostrand Co., New York, 1943.
   This text is out of print; however, many libraries may still have a copy, and used copies sometimes can be found. If you can find a copy, grab it!
3. NAGLE, R. KENT, and SAFF, EDWARD B., *Fundamentals of Differential Equations and Boundary Value Problems*, 2nd ed., Addison-Wesley Publishing Co., Boston, 1996.
4. VARBERG, DALE, and PURCELL, EDWIN J., *Calculus*, 7th ed., Prentice-Hall, Inc., Upper Saddle River, NJ, 1997.

PROBLEMS

1. Solve the following linear differential equations:

(a) $\frac{dy}{dx} + 3y = 0$

(b) $\frac{dy}{dx} - 3y = 0$

(c) $\frac{d^2y}{dx^2} + 2\frac{dy}{dx} + y = 0$

(d) $\frac{d^2y}{dx^2} - 6\frac{dy}{dx} + 9y = 0$

(e) $\frac{d^2y}{dx^2} + 9y = 0$

(f) $\frac{dx}{dt} = k_1(a - x) - k_2x$; k_1, k_2, and a are constants.

(g) $\frac{d\phi}{dr} = -a\phi$; a is constant.

(h) $\frac{d(A)}{(A)} = -k\,dt$; k is constant.

(i) $\frac{1}{\Phi(\phi)}\frac{d^2\Phi}{d\phi^2} = -m^2$; m is constant.

(j) $m\frac{d^2y}{dt^2} = -ky$; m and k are constants.

(k) $\frac{d^2\psi}{dx^2} + \frac{8\pi^2 mE}{h^2}\psi = 0$; E, m, and h are constants.

2. Test the following differentials for exactness:

(a) $dF = 2xy^2\,dx + 2yx^2\,dy$

(b) $dF = 8x\,dx$

(c) $dF = 12x^2y\,dx + 4x^3\,dy$

(d) $dF = 5\,dx$

(e) $dF = \frac{1}{y}\,dx - \frac{x}{y^2}\,dy$

(f) $dF = xy\,dx + x^3\,dy$

(g) $dP = \frac{nR}{V}\,dT - \frac{nRT}{V^2}\,dV$; n and R are constants.

(h) $dV = \pi r^2\,dh + 2\pi rh\,dr$

(i) $dq = nC_v\,dT + \frac{nRT}{V}\,dV$; n, C_v, and R are constants.

(j) $d\rho = -\frac{PM}{RT^2}\,dT + \frac{M}{RT}\,dP$; M and R are constants.

(k) $dE = nC_v\,dT + \frac{n^2 a}{V^2}\,dV$; n, C_v, and a are constants

3. Show that the differential $dq = nC_v\,dT + (nRT/V)\,dV$, where n, C_v, and R are constants, can be made exact by multiplying by an integrating factor $1/T$. The resulting differential dS is called the *differential entropy change.*

4. Show that if $\sin 3x$ and $\cos 3x$ are particular solutions to the differential equation

$$\frac{d^2y}{dx^2} + 9y = 0$$

then a linear combination of the two solutions also is a solution.

5. Bessel's equation is an important differential equation having the general form

$$x^2\frac{d^2y}{dx^2} + x\frac{dy}{dx} + (x^2 - c^2)y = 0$$

where c is a constant. Find the indicial equation for the series solution to this equation.

6. A one-dimensional harmonic oscillator is described classically by the equation

$$\frac{d^2y}{dt^2} + 4\pi^2\nu^2 y = 0$$

Show that a solution to this equation is $y = A\sin 2\pi\nu t$, where A, π, and ν are constants.

7. The differential equation describing the spacial behavior of a one-dimensional wave is

$$\frac{d^2f}{dx^2} + \frac{4\pi^2}{\lambda^2}f(x) = 0$$

where λ is the wavelength. Find the general solution to this equation.

8. *Boundary conditions* are special restrictions imposed on the solutions to differential equations. The boundary conditions for a plucked string bound at both ends between $x = 0$ and $x = L$, and described by the equation given in Problem 7, are that $f(x)$ goes to 0 at $x = 0$ and $x = L$. Show how these boundary conditions affect the solution to the equation in Problem 7.

9. Show that if we let $x = \cos\theta$, the solution to the associated Legendre's equation [Equation (6-47)] is $y = \cos\theta$, when $\ell = 1$ and $m = 0$, and is $y = \sin\theta$, when $\ell = 1$ and $m = 1$.
10. Show that if we let $x = \cos\theta$, the solution to the associated Legendre's equation [Equation (6-47)] is $y = \frac{1}{2}(3\cos^2\theta - 1)$, when $\ell = 2$ and $m = 0$.
11. The Schrödinger equation for a particle in a three-dimensional box is

$$\frac{\partial^2\psi}{\partial x^2} + \frac{\partial^2\psi}{\partial y^2} + \frac{\partial^2\psi}{\partial z^2} + \frac{2mE}{\hbar^2}\psi = 0$$

where E, m, and $\hbar$ are constants and $\psi = f(x, y, z)$. Separate the equation to an equation in x, an equation in y, and an equation in z by assuming that

$$\psi(x, y, z) = f(x)g(y)h(z)$$

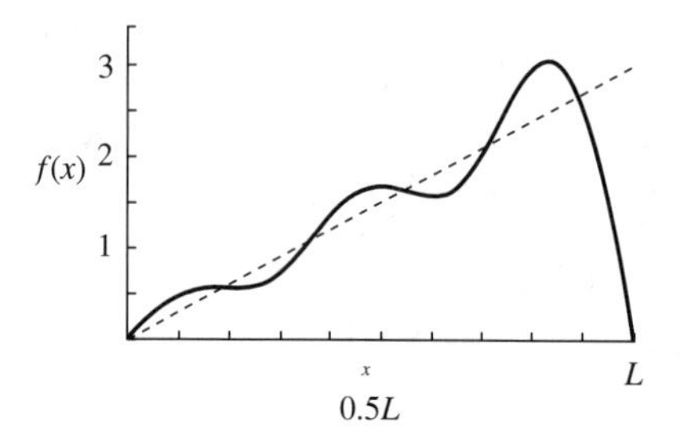

7 Infinite Series

7-1 INTRODUCTION

We saw in preceding chapters that it is sometimes useful to express a function as a sum of terms called a *series*. For example,

$$e^x = 1 + x + \frac{x^2}{2} + \frac{x^3}{6} + \frac{x^4}{24} + \cdots$$

or

$$\sin x = x - \frac{x^3}{6} + \frac{x^5}{120} - \frac{x^7}{5040} + - \cdots$$

The three dots at the end of each of the above examples signify that the number of terms in the series is endless. Therefore, such a series is called an *infinite series*.

Suppose we have a sequence of terms $u_1, u_2, u_3, \ldots$ and that we let S_n be the sum of the first n terms. That is,

$$S_n = u_1 + u_2 + u_3 + \cdots + u_n \tag{7-1}$$

If, as n approaches infinity, the sum S_n approaches a definite, finite value, then the series is said to be *convergent*. On the other hand, if, as n approaches infinity, S_n does not have a definite, finite value but increases without limit, then the series is said to be *divergent*. Several examples of convergent series and their sums are:

$$S = 1 - \frac{1}{2} + \frac{1}{3} - \frac{1}{4} + - \cdots = \sum_{x=1}^{\infty} \frac{(-1)^{x-1}}{x} = \ln 2$$

$$S = \frac{1}{2} + \frac{1}{6} + \frac{1}{12} + \frac{1}{20} + \cdots = \sum_{x=1}^{\infty} \frac{1}{x(x+1)} = 1$$

$$S = 1 + a + a^2 + a^3 + \cdots = \sum_{x=1}^{\infty} a^x = \frac{1}{1-a}; \qquad |a| < 1$$

$$S = a + 2a^2 + 3a^3 + \cdots = \sum_{x=1}^{\infty} x a^x = \frac{a}{(1-a)^2}; \qquad |a| < 1$$

7-2 TEST FOR CONVERGENCE AND DIVERGENCE

There are several ways in which to determine whether a series is convergent or divergent. Two of these methods, *comparison tests* and the *ratio test*, are outlined below. Others may be found in general calculus texts.

Comparison Tests. The first way to test whether a series converges or diverges is to compare it to one of the series, called *comparison series*, listed below. Consider the following series:

(1) The series

$$a + ar + ar^2 + ar^3 + \cdots + ar^n + \cdots$$

converges when $r < 1$ and diverges when $r \geq 1$.

(2) The series

$$1 + \frac{1}{2^p} + \frac{1}{3^p} + \frac{1}{4^p} + \cdots + \frac{1}{n^p} + \cdots$$

converges when $p > 1$.

(3) The series

$$1 + \frac{1}{2} + \frac{1}{3} + \frac{1}{4} + \cdots + \frac{1}{n} + \cdots$$

always diverges.

(4) The series

$$\frac{1}{a(a+1)} + \frac{1}{(a+1)(a+2)} + \cdots + \frac{1}{(a+n-1)(a+n)} + \cdots$$

where $a > 0$, always converges.

If each term in an unknown series is less than or equal to a corresponding term in a comparison series that is known to converge, then the unknown series converges. However, if each term in an unknown series is equal to or greater than a corresponding term in a comparison series that is known to diverge, then the unknown series diverges. In order to illustrate this method, consider the following examples:

Examples

(a) Determine whether the series

$$1 + \frac{1}{3} + \frac{1}{5} + \frac{1}{7} + \cdots$$

converges or diverges.

Solution. To solve this problem, let us compare this series with series (2) and series (3) above. If we let $p = 1$ in series (2), we have series (3)

$$1 + \frac{1}{2} + \frac{1}{3} + \frac{1}{4} + \frac{1}{5} + \cdots$$

which is divergent. However, each term in the unknown series is less than each term in this series, which seems to indicate that the series is not divergent. If we let $p = 2$ in series (2), we have

$$1 + \frac{1}{4} + \frac{1}{9} + \frac{1}{16} + \cdots$$

which is convergent. However, each term in the unknown series is not less than each term in this series. Clearly, then, there must be some value of p lying between 1 and 2 that will give a series having terms that are greater than the corresponding terms in the unknown series. Since p will be greater than 1, and this series must converge, the unknown series also must converge.

(b) Determine whether the series

$$1 + 2! + 3! + 4! + 5! + \cdots$$

converges or diverges.

Solution. To solve this problem, let us first rewrite the series as

$$1 + 2 + 6 + 24 + 120 + \cdots$$

Now, let us compare this series with series (1) in which $a = 1$ and $r = 2$.

$$1 + 2 + 4 + 8 + 16 + \cdots$$

Since each term in the known series is greater than each term in this divergent series, the unknown series must diverge.

Ratio Test. Another method used to determine whether a series is convergent or divergent is the so-called ratio test. Consider the ratio of the $(n + 1)$st term to the nth term in the infinite series $u_1 + u_2 + u_3 + \cdots$. We find that if

$$\lim_{n\to\infty}\left|\frac{u_{n+1}}{u_n}\right| < 1$$

the series converges. On the other hand, if

$$\lim_{n\to\infty}\left|\frac{u_{n+1}}{u_n}\right| > 1$$

the series diverges. If, however,

$$\lim_{n\to\infty}\left|\frac{u_{n+1}}{u_n}\right| = 1$$

the test fails and another method must be used.

Example

Determine whether the series

$$\frac{1}{2} + \frac{2}{2^2} + \frac{3}{2^3} + \frac{4}{2^4} + \cdots$$

converges or diverges.

Solution. In this series, the general term is $u_n = n/2^n$. Hence,

$$u_{n+1} = \frac{n+1}{2^{n+1}}$$

$$\lim_{n\to\infty}\left|\frac{u_{n+1}}{u_n}\right| = \lim_{n\to\infty}\left|\frac{n+1}{2^{n+1}}\cdot\frac{2^n}{n}\right| = \lim_{n\to\infty}\left|\frac{n+1}{2n}\right|$$

However, since

$$\lim_{n\to\infty}\left|\frac{n+1}{2n}\right| = \frac{1}{2}$$

and this is less than 1, the series must converge.

7-3 POWER SERIES REVISITED

One of the most useful mathematical tools in applied mathematics is the *power series*, an infinite series having the form

$$a_0 + a_1 x + a_2 x^2 + a_3 x^4 + \cdots + a_n x^n + \cdots = \sum_{n=0}^{\infty} a_n x^n$$

Power series, like other infinite series, can be either convergent or divergent; however, whether a series converges or diverges depends on the value of x.

We can determine the value of x for which a series converges or diverges by applying the ratio test. We find that the series converges for all values of x in the interval

$$|x| < \lim_{n\to\infty} \left| \frac{a_{n-1}}{a_n} \right|$$

and that the series diverges for all other values of x

$$|x| > \lim_{n\to\infty} \left| \frac{a_{n-1}}{a_n} \right|$$

At the endpoints of the interval, that is, at

$$|x| = \lim_{n\to\infty} \left| \frac{a_{n-1}}{a_n} \right|$$

the test fails.

Examples

(a) Determine for which values of x the series

$$\sum_{n=0}^{\infty} a_n x^{n+1} = a_0 x + a_1 x^2 + a_2 x^3 + \cdots = x + 2x^2 + 3x^3 + \cdots$$

converges.

Solution. In this case, $a_0 = 1$, $a_1 = 2$, $a_2 = 3$, and so on. Hence,

$$a_n = n + 1 \quad \text{and} \quad a_{n\text{-}1} = n$$

Thus,

$$\lim_{n\to\infty} \left| \frac{a_{n-1}}{a_n} \right| = \lim_{n\to\infty} \left| \frac{n}{n+1} \right| = 1$$

The series converges for $|x| < 1$, that is, in the interval $-1 < x < +1$. Likewise, the series diverges for $|x| > 1$, that is, for $x < -1$ or $x > +1$. When $x = 1$, the series becomes

$$1 + 2 + 3 + 4 + 5 + \cdots$$

which diverges. When $x = -1$, the series becomes

$$-1 + 2 - 3 + 4 - 5 + \cdots$$

which also diverges. Hence, the interval of convergence does not include the endpoints.

(b) Power series solutions to differential equations are particularly interesting when the series converges to a polynomial. For example, we can use the ratio test to determine the interval of convergence for Legendre's equation.

Solution. One will see from Equation (6-43) that

$$\left|\frac{u_{n+2}}{u_n}\right| = \left|\frac{a_{n+2}}{a_n}\right| x^2$$

But from the recursion formula for Legendre's equation, we see that

$$\lim_{n\to\infty}\left|\frac{a_{n+2}}{a_n}\right| = 1$$

Therefore, the interval of convergence for Legendre's equation is when $x^2 < 1$ or $|x| < 1$. Thus, in the interval $-1 < x < 1$, the solutions to Legendre's equation are significant. Moreover, we find that if ℓ in Legendre's equation is a positive or negative integer (including zero), the Legendre polynomials will remain finite for all the allowed values that the cosine of an angle can have (including the endpoint values $x = \cos\theta = \pm 1$).

7-4 MACLAURIN AND TAYLOR SERIES

In this chapter we consider several methods of expanding functions in infinite series. Two, which are particularly useful in physical chemistry, are the power series known as the *Maclaurin series* and the *Taylor series*. Let us consider the Maclaurin series first.

Suppose that a function $f(x)$ can be expanded in a power series

$$f(x) = a_0 + a_1x + a_2x^2 + a_3x^3 + \cdots \tag{7-2}$$

Further, suppose that the function has continuous derivatives of all orders. Let us evaluate $f(x)$ and its derivatives at $x = 0$.

$$f(x) = a_0 + a_1x + a_2x^2 + a_3x^3 + \cdots;\ f(0) = a_0$$
$$f'(x) = a_1 + 2(1)a_2x + (3)(1)a_3x^2 + \cdots;\ f'(0) = a_1$$
$$f''(x) = 2(1)a_2 + (3)(2)(1)a_3x + \cdots;\ f''(0) = 2!a_2$$
$$f^n(x) = n!a_n + (n+1)!a_{n+1}x + \cdots;\ f^n(0) = n!a_n$$

Substituting these values for the coefficients back into Equation (7-2) gives

$$f(x) = f(0) + f'(0)x + \frac{f''(0)}{2!}x^2 + \cdots + \frac{f^n(0)}{n!}x^n + \cdots \tag{7-3}$$

which is known as the Maclaurin series. To illustrate the use of this series, let us expand the function $f(x) = \sin x$ in a Maclaurin series.

$$\begin{aligned} f(x) &= \sin x & f(0) &= 0 \\ f'(x) &= \cos x & f'(0) &= 1 \\ f''(x) &= -\sin x & f''(0) &= 0 \end{aligned}$$

etc.

$$\sin x = x - \frac{x^3}{3!} + \frac{x^5}{5!} - \frac{x^7}{7!} + - \cdots$$

PROBLEM. Expand the function $\ln(1 + x)$ in a Maclaurin series.

$$\begin{aligned} f(x) &= \ln(1 + x) & f(0) &= 0 \\ f'(x) &= \frac{1}{1 + x} & f'(0) &= 1 \\ f''(x) &= \frac{-1}{(1 + x)^2} & f''(0) &= -1 \end{aligned}$$

etc.

$$\ln(1 + x) = x - \frac{x^2}{2} + \frac{x^3}{3} - + \cdots$$

Consider, now, a function $f(x)$ that can be expanded in the series

$$f(x) = c_0 + c_1(x - a) + c_2(x - a)^2 + c_3(x - a)^3 + \cdots \tag{7-4}$$

where a is some constant. Assume, as we did above, that the function $f(x)$ has continuous derivatives of all orders. Let us now evaluate this function and its derivatives at $x = a$, rather than at $x = 0$.

$$\begin{aligned} f(x) &= c_0 + c_1(x - a) + c_2(x - a)^2 + c_3(x - a)^3 + \cdots; \; f(a) = c_0 \\ f'(x) &= c_1 + 2c_2(x - a) + 3c_3(x - a)^2 + \cdots; \; f'(a) = c_1 \\ f''(x) &= 2c_2 + (3)(2)(1)c_3(x - a) + \cdots; \; f''(a) = 2!c_2 \\ f^n(x) &= n!c_n + (n + 1)!c_{n+1}(x - a) + \cdots; \; f^n(a) = n!c_n \end{aligned}$$

Substituting these values for the coefficients back into Equation (7-4) gives

$$f(x) = f(a) + f'(a)(x-a) + \frac{f''(a)}{2!}(x-a)^2 + \cdots + \frac{f^n(a)}{n!}(x-a)^n + \cdots \tag{7-5}$$

which is known as a Taylor's series.

PROBLEM. Expand the function $f(x) = e^x$ in powers of $(x+2)$. Since $f(x) = e^x$ and $(x+2) = (x-a)$ or $a = -2$, we have

$$f(x) = e^x \qquad f(a) = e^{-2}$$

$$f'(x) = e^x \qquad f'(a) = e^{-2}$$

$$f''(x) = e^x \qquad f''(a) = e^{-2}$$

etc.

$$f(x) = e^{-2}\left[1 + (x+2) + \frac{1}{2}(x+2)^2 + \frac{1}{6}(x+2)^3 + \cdots\right]$$

7-5 FOURIER SERIES AND FOURIER TRANSFORMS

One of the most common mathematical tools in chemistry and physics is the Fourier transform. Fourier transforms are mathematical manipulations that reorganize information. They arise naturally in a number of physical problems. For example, a lens is a Fourier transformer. The human eye and ear act as Fourier transformers by analyzing complex electromagnetic or sound waves.

A Fourier transform is performed with the use of a Fourier series, expressed in its most general form as the sum of sine and cosine functions

$$f(x) = \sum_n a_n \sin nx + \frac{1}{2}b_0 + \sum_n b_n \cos nx \tag{7-6}$$

where

$$a_n = \frac{1}{\pi}\int_{-\pi}^{\pi} f(x)\sin nx\,dx \quad \text{and} \quad b_n = \frac{1}{\pi}\int_{-\pi}^{\pi} f(x)\cos nx\,dx \tag{7-7}$$

The coefficients are found from the *orthonormal* behavior of sine and cosine functions, which we will discuss later in the chapter. Fourier series differ from power series expansions in at least two important ways. First, the interval of convergence of a power series is different for different functions; the Fourier series, on the other hand, always converges between $-\pi$ and $+\pi$. Second, many functions cannot be expanded in a power series, whereas it is rare to find a function that cannot be expanded in a Fourier series.

It is sometimes useful to change the range of the Fourier series from $(-\pi, \pi)$ to $(-L, L)$. This is accomplished by replacing the variable x in Equation (7-6) with

$\pi x/L$, giving

$$f(x) = \sum_n a_n \sin\frac{n\pi x}{L} + \frac{1}{2}b_0 + \sum_n b_n \cos\frac{n\pi x}{L} \tag{7-8}$$

where now

$$a_n = \frac{1}{L}\int_{-L}^{+L} f(x)\sin\frac{n\pi x}{L}dx \quad \text{and} \quad b_n = \frac{1}{L}\int_{-L}^{+L} f(x)\cos\frac{n\pi x}{L}dx \tag{7-9}$$

We saw in Chapter 1 that a sum of sine and cosine functions also can be represented as a complex exponential. It is easy to show, therefore, that another form of the Fourier series is

$$f(x) = \sum_{-\infty}^{+\infty} c_n e^{in\pi x/L} \tag{7-10}$$

where $c_n = \frac{1}{2L}\int_{-L}^{+L} f(x)e^{-in\pi x/L}dx$. Note that, in this case, the coefficients are complex. This series, however, will represent only functions that are periodic. It is possible, though, to modify Equation (7-10) to represent functions that are nonperiodic. Let $k = n\pi/L$. Now, let us see what happens if we allow L to go to infinity. As L gets larger and larger, k changes in smaller and smaller increments with each change in n: $\Delta k = (\pi/L)\,\Delta n$. In the limit that Δk goes to zero, k becomes a continuous variable and the coefficients c_k can be described as a function of k, $c(k)$. Therefore,

$$f(x) = \lim_{L\to\infty}\sum_{-\infty}^{+\infty} c_n e^{in\pi x/L}\,\Delta n$$

$$f(x) = \lim_{\Delta k\to 0}\sum_{-\infty}^{+\infty} \frac{Lc(k)}{\pi}e^{ikx}\,\Delta k$$

But this is just the definition of the integral. Letting $Lc(k)/\pi = g(k)/\sqrt{2\pi}$, we have

$$f(x) = \frac{1}{\sqrt{2\pi}}\int_{-\infty}^{+\infty} g(k)e^{ikx}\,dk \tag{7-11}$$

and

$$g(k) = \frac{1}{\sqrt{2\pi}}\int_{-\infty}^{+\infty} f(x)e^{-ikx}\,dx \tag{7-12}$$

Equation (7-11) is called the *Fourier integral*, and $f(x)$ and $g(k)$ are called *Fourier transforms* of one another.

PROBLEM. Find the Fourier transform of the function

$$f(x) = \begin{cases} 0; & x < -nL \\ \sin\dfrac{2\pi x}{L}; & -nL < x < nL \\ 0; & x > nL \end{cases}$$

The Fourier transform of this function is

$$g(k) = \frac{1}{\sqrt{2\pi}} \int_{-nL}^{+nL} \sin \frac{2\pi x}{L} e^{-ikx}\, dx$$

Note that since the function equals zero outside the range $(-nL, nL)$, we need only to integrate from $-nL$ to $+\,nL$, rather than from $-\infty$ to $+\infty$. If we let $a = -ik$ and $b = 2\pi/L$, this integral becomes $\int e^{ax} \sin bx\, dx$, which can be found in the Table of Integrals in Appendix II. Therefore, we have

$$g(k) = \frac{1}{\sqrt{2\pi}} \left(\frac{1}{a^2 + b^2}\right) \left[e^{ax} (a \sin bx - b \cos bx)\right]_{-nL}^{+nL}$$

$$g(k) = \frac{1}{\sqrt{2\pi}} \left(\frac{1}{a^2 + b^2}\right) \left[e^{-ikx} \left(-ik \sin \frac{2\pi x}{L} - \left(\frac{2\pi}{L}\right) \cos \frac{2\pi x}{L}\right)\right]_{-nL}^{+nL}$$

Substituting the limits of integration into the integral, we have

$$g(k) = \frac{1}{\sqrt{2\pi}} \left(\frac{1}{a^2 + b^2}\right) \left[e^{-iknL} \left(-ik \sin 2\pi n - \left(\frac{2\pi}{L}\right) \cos 2\pi n\right) - e^{iknL} \left(-ik \sin(-2\pi n) - \left(\frac{2\pi}{L}\right) \cos(-2\pi n)\right)\right]$$

But $\sin(2\pi n) = 0 = \sin(-2\pi n)$ and $\cos(2\pi n) = \cos(-2\pi n) = 1$. Therefore, we can write

$$g(k) = \frac{1}{\sqrt{2\pi}} \left(\frac{1}{a^2 + b^2}\right) \left[-e^{-iknL} \left(\frac{2\pi}{L}\right) + e^{+iknL} \left(\frac{2\pi}{L}\right)\right]$$

$$g(k) = \frac{1}{\sqrt{2\pi}} \left(\frac{1}{a^2 + b^2}\right) \left(\frac{2\pi}{L}\right) [-e^{-iknL} + e^{+iknL}]$$

From Chapter 1, $[e^{iknL} - e^{-iknL}] = 2i \sin(nkL)$. Also $a^2 + b^2 = \frac{4\pi^2}{L^2} - k^2 = \frac{4\pi^2 - k^2L^2}{L^2}$. Substituting these into the equation for $g(k)$ gives

$$g(k) = \frac{2i\sqrt{2\pi}\, L}{4\pi^2 - k^2 L^2} \sin(nkL)$$

Simpler forms of the Fourier series are possible. For example, if the function to be expanded is an *odd* function about $x = 0$, that is, $f(x) = -f(-x)$, then the b coefficients in Equation (7-6) vanish and the Fourier series becomes

$$f(x) = \sum_{n=1}^{\infty} a_n \sin nx \tag{7-13}$$

where

$$a_n = \frac{2}{\pi} \int_0^{\pi} f(x) \sin nx \, dx \tag{7-14}$$

Likewise, if the function $f(x)$ is an *even* function about $x = 0$, that is, $f(x) = f(-x)$, then the a coefficients in Equation (7-6) vanish and the Fourier series becomes

$$f(x) = \frac{1}{2} b_o + \sum_{n=1}^{\infty} b_n \cos nx \tag{7-15}$$

where

$$b_n = \frac{2}{\pi} \int_0^{\pi} f(x) \cos nx \, dx \tag{7-16}$$

Consider, now, another set of functions that can be used as the basis set for a Fourier expansion. The wave-mechanical solutions to a particle confined to a one-dimensional box between $x = 0$ and $x = L$ are

$$\psi_n = \sqrt{\frac{2}{L}} \sin \frac{n\pi x}{L} \tag{7-17}$$

where n is a positive integer (not including zero). These solutions have at least two very important properties. First, the solutions are said to be *normalized*. The condition for normalization is

$$\int_{\text{all space}} \psi_i^* \psi_i \, d\tau = 1 \tag{7-18}$$

where ψ^* denotes the complex conjugate of ψ and $d\tau$ is the differential volume element. Second, any two different solutions are said to be *orthogonal*. The condition for orthogonality is

$$\int_{\text{all space}} \psi_i^* \psi_j \, d\tau = 0 \tag{7-19}$$

Taken together, these conditions are said to form a *complete orthonormal set*, sometimes designated as

$$\int_{\text{all space}} \psi_i^* \psi_j \, d\tau = \delta_{ij}; \quad \delta_{ij} = \begin{cases} 1, & i = j \\ 0, & i \neq j \end{cases} \tag{7-20}$$

The term δ_{ij} is called *Kronecker delta*.

We can use this complete set as the basis set for a Fourier series expansion of any function in the interval from 0 to L. For example, let us expand the function $f(x) = 3x$ in a Fourier series, using the solutions of the particle in the box as the basis set.

$$f(x) = \sum_n a_n \psi_n = \sum_n a_n \sqrt{\frac{2}{L}} \sin \frac{n\pi x}{L} \tag{7-21}$$

To find the coefficients, we make use of the orthonormal properties of ψ. Let us multiply through Equation (7-21) by ψ^* and integrate over all space (in this case from 0 to L).

$$\int_0^L \psi_m^* f(x)\,dx = \int_0^L \psi_m^* \sum_n a_n \psi_n\,dx = \int_0^L \sum_n a_n \psi_m^* \psi_n\,dx$$

But the integral of a sum is the sum of the integrals.

$$\int_0^L \sum_n a_n \psi_m^* \psi_n\,dx = \sum_n a_n \int_0^L \psi_m^* \psi_n\,dx$$

Every integral in this summation will vanish, except when $n = m$, in which case the integral is equal to 1. Thus, we can write

$$a_m = \int_0^L \psi_m^* f(x)\,dx = \sqrt{\frac{2}{L}} \int_0^L f(x) \sin\frac{m\pi x}{L}\,dx \tag{7-22}$$

For the function $f(x) = 3x$, Equation (7-22) is

$$a_m = 3\sqrt{\frac{2}{L}} \int_0^L x \sin\frac{m\pi x}{L}\,dx$$

which integrates to give

$$a_n = -3\sqrt{\frac{2}{L}}\left(\frac{L^2}{n\pi}\right)(-1)^n$$

Using these coefficients, we now expand the function in a Fourier series

$$f(x) = \sum_n -\frac{6L}{n\pi}(-1)^n \sin\frac{n\pi x}{L}$$

$$f(x) = \frac{6L}{\pi}\left(\sin\frac{\pi x}{L} - \frac{1}{2}\sin\frac{2\pi x}{L} + \frac{1}{3}\sin\frac{3\pi x}{L} - \cdots\right)$$

A graph of the first five terms of this series, along with a graph of the original function (dashed line) between $x = 0$ and $x = L$, is shown in Fig. 7-1. (Since the function is plotted in fractional units of L, the graph actually is from $x = 0$ to $x = 1$.) Note that even with only five terms in the Fourier series, the series does a pretty fair job of following the actual function. Of course, more terms in the series will give a better fit. The series does struggle to fit the function as x approaches L, since the actual function is discontinuous at that point, while the series goes to zero. This, however, is normal when one attempts to approximate a nonperiodic, discontinuous function with a sum of sine waves.

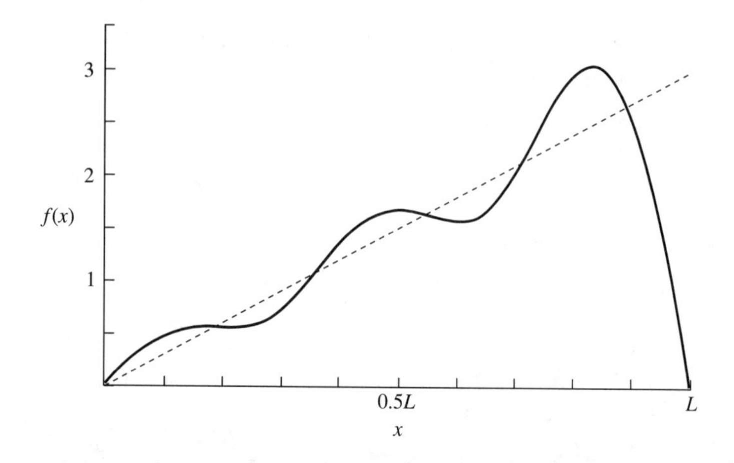

Figure 7-1 Graph of first five terms of the Fourier series expansion of the function $f(x) = 3x$, along with graph of the actual function (dashed line).

SUGGESTED READING

1. Bradley, Gerald L., and Smith, Karl J., *Calculus*, Prentice-Hall, Inc., Upper Saddle River, NJ, 1995.
2. Nagle, R. Kent, and Saff, Edward B., *Fundamentals of Differential Equations and Boundary Value Problems*, 2nd ed., Addison-Wesley Publishing Co., Boston, 1996.
3. Varberg, Dale, and Purcell, Edwin J., *Calculus*, 7th ed., Prentice-Hall, Inc., Upper Saddle River, NJ, 1997.

PROBLEMS

1. Using the comparison tests, determine whether each of the following series are convergent or divergent:

   (a) $1 + 3 + 5 + 7 + 9 + \cdots$

   (b) $\frac{3}{2} + \frac{3}{4} + \frac{3}{6} + \frac{3}{8} + \frac{3}{10} + \cdots$

   (c) $\frac{1}{1(2)} + \frac{1}{2(3)} + \frac{1}{3(4)} + \frac{1}{4(5)} + \cdots$

   (d) $1 + \frac{1}{3} + \frac{1}{6} + \frac{1}{12} + \frac{1}{24} + \cdots$

   (e) $1 + \frac{1}{2!} + \frac{1}{3!} + \frac{1}{4!} + \frac{1}{5!} + \cdots$

   (f) $\frac{2}{1} + \frac{3}{2} + \frac{4}{3} + \frac{5}{4} + \frac{6}{5} + \cdots$

(g) $1 + \frac{3}{2} + \frac{9}{4} + \frac{27}{8} + \frac{81}{16} + \cdots$

(h) $\frac{1}{4} + \frac{1}{12} + \frac{1}{36} + \frac{1}{108} + \frac{1}{324} + \cdots$

(i) $\frac{1}{2} + \frac{2!}{2^2} + \frac{3!}{2^3} + \frac{4!}{2^4} + \cdots$

(j) $1 + \frac{1}{3} + \frac{1}{4^2} + \frac{1}{5^3} + \frac{1}{6^4} + \cdots$

2. Using the ratio test, determine whether the following series are convergent or divergent:

(a) $\frac{1}{2} + \frac{1}{2^2} + \frac{1}{2^3} + \frac{1}{2^4} + \frac{1}{2^5} + \cdots$

(b) $3 + \frac{3^2}{2} + \frac{3^3}{3} + \frac{3^4}{4} + \frac{3^5}{5} + \cdots$

(c) $\frac{1}{2} + \frac{2}{3} + \frac{3}{4} + \frac{4}{5} + \frac{5}{6} + \cdots$

(d) $2 + \frac{2^2}{2^2} + \frac{2^3}{3^2} + \frac{2^4}{4^2} + \frac{2^5}{5^2} + \cdots$

(e) $\frac{1}{2} + \frac{2^2}{2^2} + \frac{3^2}{2^3} + \frac{4^2}{2^4} + \frac{5^2}{2^5} + \cdots$

(f) $3 + \frac{3^2}{2!} + \frac{3^3}{3!} + \frac{3^4}{4!} + \frac{3^5}{5!} + \cdots$

(g) $1 + \frac{1!}{2} + \frac{2!}{3} + \frac{3!}{4} + \frac{4!}{5} + \cdots$

(h) $\frac{1}{3!} + \frac{1}{6!} + \frac{1}{9!} + \frac{1}{12!} + \frac{1}{15!} + \cdots$

(i) $1 + \frac{1}{2} + \frac{2!}{2^2} + \frac{3!}{2^3} + \frac{4!}{2^4} + \cdots$

(j) $\frac{1}{3} + \frac{2^2}{4} + \frac{3^2}{5} + \frac{4^2}{6} + \frac{5^2}{7} + \cdots$

3. Determine the interval of convergence for the following power series:

(a) $1 + x + x^2 + x^3 + \cdots$

(b) $1 - 2x + 3x^2 - 4x^3 + - \cdots$

(c) $1 + x + \frac{x^2}{2!} + \frac{x^3}{3!} + \cdots$

(d) $x - \frac{x^2}{2} + \frac{x^3}{3} - \frac{x^4}{4} + - \cdots$

(e) $1 - x^2 + \frac{x^4}{2!} - \frac{x^6}{3!} + - \cdots$

(f) $x - \frac{1}{3}x^3 + \frac{1}{5}x^5 - \frac{1}{7}x^7 + - \cdots$

(g) $x - \frac{x^3}{3!} + \frac{x^5}{5!} - \frac{x^7}{7!} + - \cdots$

(h) $(x-1) - \frac{1}{2}(x-1)^2 + \frac{1}{3}(x-1)^3 - \frac{1}{4}(x-1)^4 + - \cdots$

(i) $1 + \frac{x}{2} + \frac{x^2}{4} + \frac{x^3}{8} + \frac{x^4}{16} + \cdots$

(j) $1 + (x+2) + (x+2)^2 + (x+2)^3 + \cdots$

4. Expand the following functions in a Maclaurin series:

(a) $\frac{1}{1+x}$

(b) $\frac{1}{(1-x)^2}$

(c) $(1+x)^{1/2}$

(d) $\ln(1-x)$

(e) e^{-x^2}

(f) a^x

(g) $\cos x$

(h) $(x+1)^3$

5. Show that, for small values of X_B, $\ln(1–X_B) \cong -X_B$.

6. Show that, for small values of θ, $\sin\theta \cong \theta$.

7. Show, by expanding $\sin x$ in powers of $(x-a)$, that the series converges most rapidly as x approaches a.

8. Evaluate the integral $\int_0^4 e^{-x^2}dx$ by expanding the function in a Maclaurin series (first 8 terms).

9. Show that the solutions to the particle in the one-dimensional box, $\psi_n = \sqrt{\frac{2}{L}}\sin\frac{n\pi x}{L}$, are orthogonal and normalized.

10. Find the Fourier transform of the function

$$f(x) = \begin{cases} 0; & x < -\pi \\ x; & -\pi < x < \pi \\ 0; & x > \pi \end{cases}$$

11. Find the Fourier transform of the step function

$$f(x) = \begin{cases} 0; & x < -L \\ \frac{\sqrt{2\pi}}{2L}; & -L < x < L \\ 0; & x > L \end{cases}$$

12. Of the Fourier transforms that occur throughout mathematics, chemistry, and physics, perhaps the most striking are those that occur in the areas of diffraction. For example, we can use a Fourier transform to reorganize the information found in an X-ray diffraction pattern and retransform it back into an "image" of a crystal. Consider the following one-dimensional crystal structure problem:

$$F_0 = +52.0 \qquad F_3 = +25.8$$

$$F_1 = -20.0 \qquad F_4 = -8.9$$

$$F_2 = -14.5 \qquad F_5 = -7.2$$

For a centrosymmetric wave (that is, a wave that is symmetric about the region of space in which it exists), the Fourier series is

$$f(x) = F_0 + \sum_n F_n \cos 2\pi nx \qquad (n = 1 \text{ to } \infty)$$

where F_i represents the Fourier coefficients given above. Plot $f(x)$ from $x = 0$ to $x = 1$ in steps of 0.05 and show that the first six terms approximate a one-dimensional unit cell containing two atoms, one at $x = 1/3$ and the other at $x = 2/3$.

13. The step function

$$f(x) = \begin{cases} -1; & -\pi \leq x < 0 \\ +1; & 0 < x \leq \pi \end{cases}$$

can be described by the Fourier series $f(x) = \sum_n a_n \sin nx$, where

$$a_n = \frac{2}{\pi} \int_0^{\pi} f(x) \sin nx \, dx$$

Plot the actual function from $-\pi$ to $+\pi$ and compare it to a plot of the Fourier series containing the first five nonzero terms.

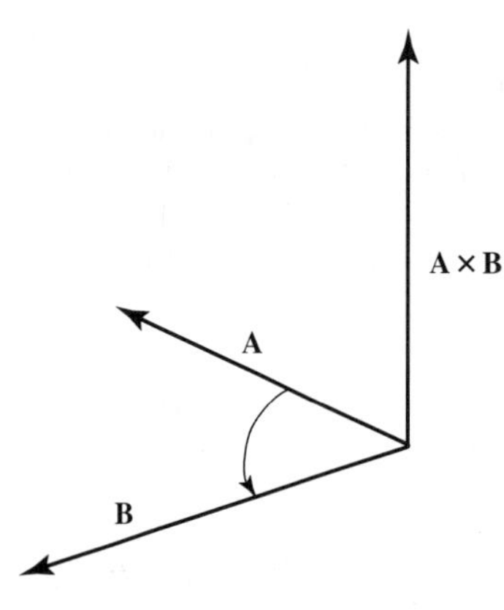

8

Scalars and Vectors

8-1 SCALARS

A *scalar* is defined as a quantity that remains invariant under a coordinate transformation. For example, consider the two points (4, 5) and (1, 2) located on the coordinate system shown in Fig. 8-1(a). The distance between the two points, S, can be found using the equation

$$S = \sqrt{(x_2 - x_1)^2 + (y_2 - y_1)^2} = \sqrt{3^2 + 3^2} = 3\sqrt{2}$$

Suppose, now, we allow the coordinate axes to be rotated about the origin through an angle of 45°, giving the new coordinate system shown in Fig. 8-1(b). The coordinates of the two points in this new coordinate system are $(-\sqrt{2}/2, 9\sqrt{2}/2)$ and $(-\sqrt{2}/2, 3\sqrt{2}/2)$. (See Chapter 10, Section 5.) The distance between the two points can be found by again using the preceding equation; thus,

$$S' = \sqrt{(x_2' - x_1')^2 + (y_2' - y_1')^2} = \sqrt{0 + (-3\sqrt{2})^2} = 3\sqrt{2}$$

Note that although the coordinates of the points changed under the coordinate transformation, the distance between the two points remained unchanged. Thus, the distance between two points is a scalar quantity. Other examples of scalars are mass, temperature, and speed (not velocity).

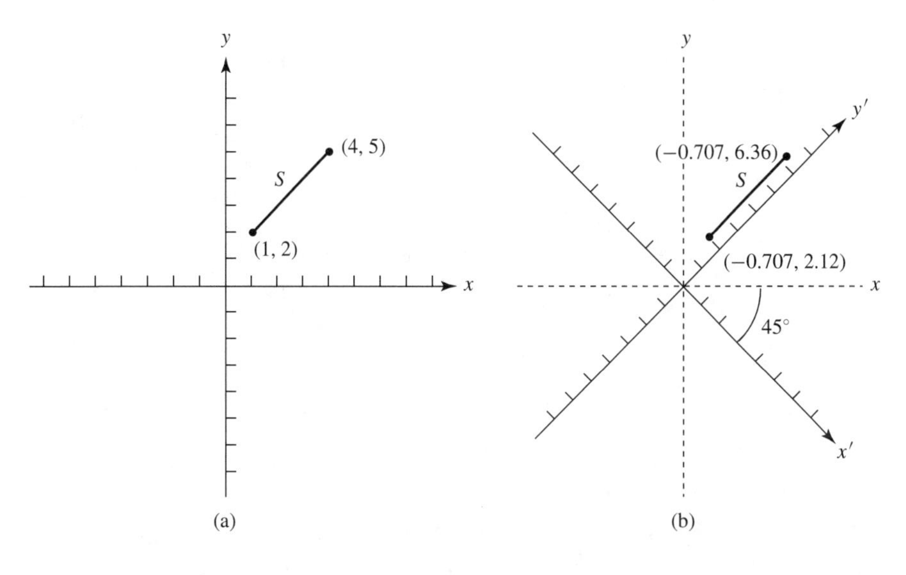

Figure 8-1 Invariance of scalar under coordinate transformation.

8-2 VECTORS AND THEIR ADDITION

Quantities that do not remain invariant under a coordinate transformation are called *vectors*. A vector is defined as a quantity having both magnitude and direction. Thus, velocity, which is a vector, has a magnitude, known as speed (10 m/s), and a direction (10 m/s along the x-axis). This is easily verified when one considers that to change from going, say, 20 miles per hour in a northerly direction to 20 miles per hour in a westerly direction (constant speed), a force must be applied to the wheels of the car. Since a force is always associated with an acceleration, and an acceleration is a change in velocity with respect to time, the velocity of the car must change even though the speed is constant.

Since vectors are quantities having both magnitude and direction, the sum of two or more vectors also must include these properties. One method of adding vectors is to use the so-called parallelogram rule. Consider two vectors **A** and **B**, represented by arrows on the coordinate system shown in Fig. 8-2(a). It is customary to let the length of the arrow represent the magnitude or absolute value of the vector (a scalar quantity) and the direction of the arrow to represent the vector's direction. To add **A** and **B**, we simply construct a parallelogram, as shown in Fig. 8-2(b). The sum of the two vectors **A** and **B**, then, is the length and direction of the diagonal of the parallelogram.

Unit Vectors. Once we know how to add vectors, we find it much more useful to represent a vector in terms of *unit vectors* lying along the axes of the coordinate system. (It is not necessary that the unit vectors lie along the coordinate axes in all coordinate systems.) Let us define **i**, **j**, and **k** as unit vectors, vectors having a

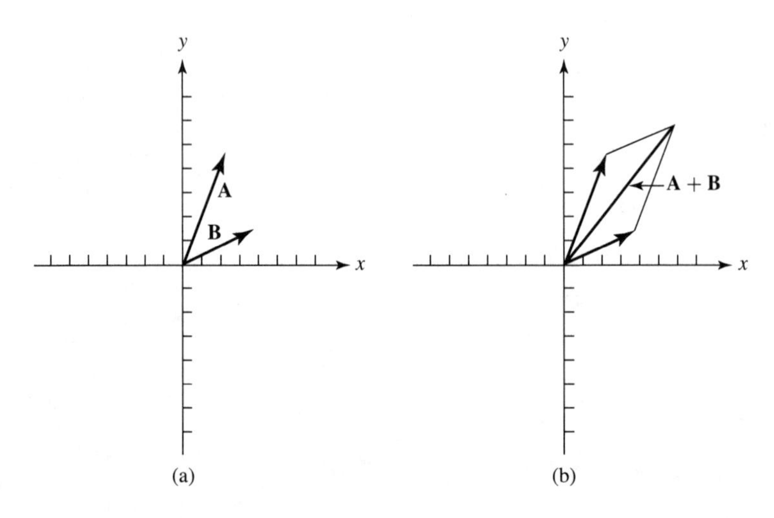

Figure 8-2 Vector addition by the parallelogram method.

magnitude (absolute value) of unity, lying along the x-, y-, and z-axes of a Cartesian coordinate system.[1] Using the method of addition described above, we find that any vector **A** can be described as the sum of multiples of these unit lengths, **i**, **j**, and **k**. Thus, we can write

$$\mathbf{A} = \mathbf{i}a_x + \mathbf{j}a_y + \mathbf{k}a_z \tag{8-1}$$

where a_x, a_y, and a_z are the scalar multiples. In terms of the scalar multiples, the magnitude or absolute value of the vector and its direction can be found using the transformation equations for plane polar coordinates [Equations (1-5) and (1-6)], if the vector is confined to the x-y plane, or the transformation equations for spherical polar coordinates [Equations (1-9), (1-10), and (1-11)], if the vector lies in three dimensions. In three dimensions we have

$$|A| = (a_x^2 + a_y^2 + a_z^2)^{1/2}$$

$$\begin{aligned} \theta &= \cos^{-1} \frac{a_z}{(a_x^2+a_y^2+a_z^2)^{1/2}} \text{ from the } z\text{-axis} \\ \phi &= \tan^{-1}\left(\frac{a_y}{a_x}\right) \text{ from the } x\text{-}z \text{ plane} \end{aligned} \tag{8-2}$$

Example

Determine the magnitude and direction of the vector $\mathbf{A}(3, 4, 3) = \mathbf{i}(3) + \mathbf{j}(4) + \mathbf{k}(3)$, illustrated in Fig. 8-3.

[1]Some texts designate the unit vectors as $\hat{\imath}$, $\hat{\jmath}$, and $\hat{k}$ (read "i-hat," "j-hat," and "k-hat").

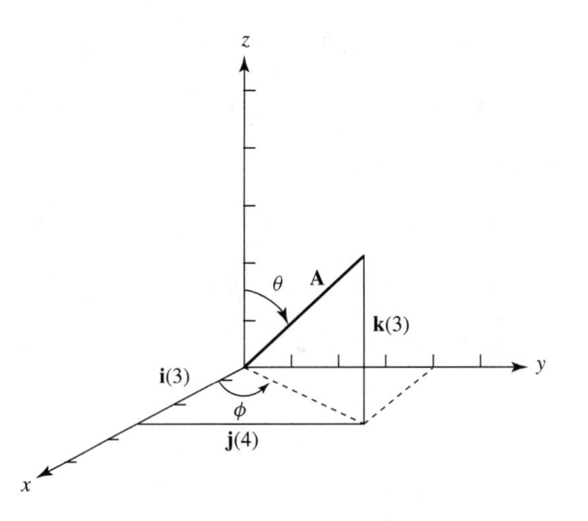

Figure 8-3 Vector $\mathbf{A}(3, 4, 3) = \mathbf{i}(3) + \mathbf{j}(4) + \mathbf{k}(3)$.

Solution:

$$|A| = \sqrt{3^2 + 4^2 + 3^2} = \sqrt{34}$$

$$\theta = \cos^{-1} \frac{3}{\sqrt{34}} = 59.04^\circ \text{ from } z\text{-axis}$$

$$\phi = \tan^{-1} \left(\frac{4}{3}\right) = 53.13^\circ \text{ from } x\text{-}z \text{ plane}$$

To consider the sum of two vectors by a nongraphical method, we must consider the component vectors along each axis separately. Consider, as an example, two vectors **A** and **B** lying in the x-y plane

$$\mathbf{A} = \mathbf{i}(1) + \mathbf{j}(2) \quad \text{and} \quad \mathbf{B} = \mathbf{i}(2) + \mathbf{j}(1)$$

The sum of the two vectors is simply the algebraic sum of the scalar multiples along each component axis. That is,

$$\begin{aligned} \mathbf{A} + \mathbf{B} &= \mathbf{i}(a_x + b_x) + \mathbf{j}(a_y + b_y) \\ &= \mathbf{i}(1 + 2) + \mathbf{j}(2 + 1) \\ &= \mathbf{i}(3) + \mathbf{j}(3) \end{aligned}$$

We see that, indeed, $\mathbf{A} + \mathbf{B}$ is a new vector $\mathbf{C} = \mathbf{i}(3) + \mathbf{j}(3)$. To find the magnitude and direction of **C**, we use the transformation equations for plane polar coordinates

$$|C| = \sqrt{3^2 + 3^2} = 3\sqrt{2}$$

$$\theta = \tan^{-1} \left(\frac{3}{3}\right) = 45^\circ \text{ from the } x\text{-axis}$$

A general expression for the addition of vectors in two dimensions is, therefore,

$$\mathbf{A}_1 + \mathbf{A}_2 + \mathbf{A}_3 + \cdots = \mathbf{i}\sum a_x + \mathbf{j}\sum a_y \tag{8-3}$$

and for three dimensions it is

$$\mathbf{A}_1 + \mathbf{A}_2 + \mathbf{A}_3 + \cdots = \mathbf{i}\sum a_x + \mathbf{j}\sum a_y + \mathbf{k}\sum a_z \tag{8-4}$$

Example

Find the sum of the three vectors $\mathbf{A}(1, 1, 2)$, $\mathbf{B}(-1, 2, -3)$, and $\mathbf{C}(2, -1, 0)$.

Solution.

$$\mathbf{A} = \mathbf{i}(1) + \mathbf{j}(1) + \mathbf{k}(2)$$
$$\mathbf{B} = \mathbf{i}(-1) + \mathbf{j}(2) + \mathbf{k}(-3)$$
$$\mathbf{C} = \mathbf{i}(2) + \mathbf{j}(-1)$$

$$\sum a_x = 1 - 1 + 2 = 2$$
$$\sum a_y = 1 + 2 - 1 = 2$$
$$\sum a_z = 2 - 3 = -1$$

Thus,

$$\mathbf{A} + \mathbf{B} + \mathbf{C} = \mathbf{i}(2) + \mathbf{j}(2) + \mathbf{k}(-1)$$

$$\text{Magnitude} = \sqrt{2^2 + 2^2 + (-1)^2} = 3$$

$$\theta = \cos^{-1}\left(\frac{-1}{3}\right) = 109.47^\circ \text{ from } z\text{-axis}$$

$$\phi = \tan^{-1}\left(\frac{2}{2}\right) = 45^\circ \text{ from } x\text{-}y \text{ plane}$$

8-3 MULTIPLICATION OF VECTORS

There are at least two ways to multiply vectors. The first way is to find the scalar or "dot" product between the two vectors. The second way is to find the vector, or "cross" product, between the two vectors. We shall consider scalar multiplication first.

Scalar Multiplication. The scalar, or "dot" product, is defined by the equation

$$\mathbf{A} \cdot \mathbf{B} = |A||B| \cos \theta_{AB} \tag{8-5}$$

where θ_{AB} is the angle between **A** and **B**. This product (read "*A* dot *B*") is called the scalar product because the result of this multiplication yields a scalar. To see this, consider first the scalar products between the unit vectors **i**, **j**, and **k**. Since the angle between any two unit vectors in Cartesian coordinates is 90°, and the magnitude of the vectors is unity, we can write

$$\begin{aligned} \mathbf{i}\cdot\mathbf{i}&=1 & \mathbf{j}\cdot\mathbf{j}&=1 \\ \mathbf{i}\cdot\mathbf{j}&=0 & \mathbf{j}\cdot\mathbf{k}&=0 \\ \mathbf{i}\cdot\mathbf{k}&=0 & \mathbf{k}\cdot\mathbf{k}&=1 \end{aligned} \tag{8-6}$$

Such vectors obeying Equation (8-6) are said to be *orthogonal.* In general, we can state that if two vectors $\mathbf{q}_i$ and $\mathbf{q}_j$ are orthogonal unit vectors, then

$$\mathbf{q}_i \cdot \mathbf{q}_j = \delta_{ij} \quad \text{where } \delta_{ij} = \begin{cases} 1 \text{ for } i = j \\ 0 \text{ for } i \neq j \end{cases} \tag{8-7}$$

We recognize δ_{ij} as *Kronecker delta* (see Chapter 7).

Consider, now, the scalar product between two vectors $\mathbf{A} = \mathbf{i}a_x + \mathbf{j}a_y + \mathbf{k}a_z$ and $\mathbf{B} = \mathbf{i}b_x + \mathbf{j}b_y + \mathbf{k}b_z$.

$$\begin{aligned} \mathbf{A}\cdot\mathbf{B} &= (\mathbf{i}a_x + \mathbf{j}a_y + \mathbf{k}a_z)\cdot(\mathbf{i}b_x + \mathbf{j}b_y + \mathbf{k}b_z) \\ &= a_xb_x\mathbf{i}\cdot\mathbf{i} + a_xb_y\mathbf{i}\cdot\mathbf{j} + a_xb_z\mathbf{i}\cdot\mathbf{k} + a_yb_x\mathbf{j}\cdot\mathbf{i} + a_yb_y\mathbf{j}\cdot\mathbf{j} + a_yb_z\mathbf{j}\cdot\mathbf{k} \\ &\quad + a_zb_x\mathbf{k}\cdot\mathbf{i} + a_zb_y\mathbf{k}\cdot\mathbf{j} + a_zb_z\mathbf{k}\cdot\mathbf{k} \end{aligned}$$

Taking into account Equations (8-6), we have

$$\mathbf{A}\cdot\mathbf{B} = a_xb_x + a_yb_y + a_zb_z \tag{8-8}$$

which, indeed, is a scalar.

Example

Find the scalar product between the vectors $\mathbf{A}(1, 3, 2)$ and $\mathbf{B}(4, -4, 1)$.

Solution.

$$\mathbf{A} = \mathbf{i}(1) + \mathbf{j}(3) + \mathbf{k}(2)$$

$$\mathbf{B} = \mathbf{i}(4) + \mathbf{j}(-4) + \mathbf{k}(1)$$

$$\mathbf{A}\cdot\mathbf{B} = 1(4) + 3(-4) + 2(1) = -6$$

Vector Multiplication. The vector or "cross" product is defined by the equation

$$\mathbf{A}\times\mathbf{B} = |A||B|\mathbf{C}\sin\theta_{AB} \tag{8-9}$$

where θ_{AB} is the angle between **A** and **B**, and **C** is a unit vector perpendicular to the plane formed by **A** and **B**. This product is called the vector product because the result of the multiplication is a vector. To obtain the direction of **A** × **B** (read "*A* cross *B*"), the "right hand rule" can be used. This rule states that if the fingers of the right hand

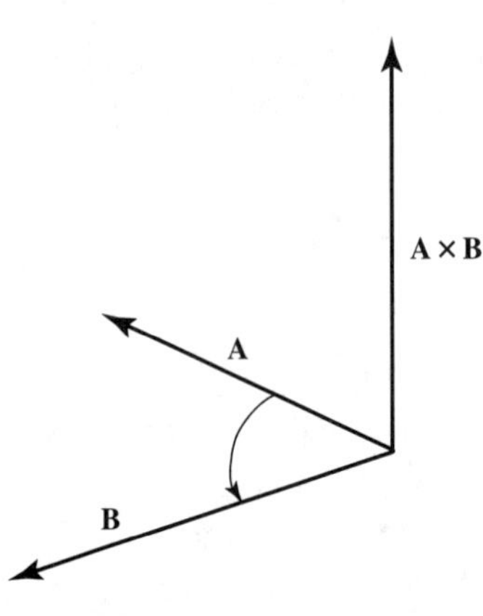

Figure 8-4 Vector product **A** × **B** perpendicular to the plane formed by vectors **A** and **B**.

rotate **A** into **B** through the smaller angle between their positive senses, the thumb will point in the direction of the cross product, shown in Fig. 8-4.

With this in mind, let us determine the vector product between the unit vectors **i**, **j**, and **k**.

$$\begin{aligned} &\mathbf{i} \times \mathbf{i} = 0 \qquad \mathbf{i} \times \mathbf{j} = \mathbf{k} \qquad \mathbf{j} \times \mathbf{j} = 0 \\ &\mathbf{j} \times \mathbf{k} = \mathbf{i} \qquad \mathbf{k} \times \mathbf{k} = 0 \qquad \mathbf{k} \times \mathbf{i} = \mathbf{j} \end{aligned} \tag{8-10}$$

From these, we can obtain an expression for the vector product between vectors **A** and **B**.

$$\begin{aligned} \mathbf{A} \times \mathbf{B} &= (\mathbf{i}a_x + \mathbf{j}a_y + \mathbf{k}a_z) \times (\mathbf{i}b_x + \mathbf{j}b_y + \mathbf{k}b_z) \\ &= a_xb_x\mathbf{i} \times \mathbf{i} + a_xb_y\mathbf{i} \times \mathbf{j} + a_xb_z\mathbf{i} \times \mathbf{k} + a_yb_x\mathbf{j} \times \mathbf{i} + a_yb_y\mathbf{j} \times \mathbf{j} \\ &\quad + a_yb_z\mathbf{j} \times \mathbf{k} + a_zb_x\mathbf{k} \times \mathbf{i} + a_zb_y\mathbf{k} \times \mathbf{j} + a_zb_z\mathbf{k} \times \mathbf{k} \end{aligned}$$

Taking into account Equations (8-10), we have

$$\mathbf{A} \times \mathbf{B} = \mathbf{i}(a_yb_z - a_zb_y) + \mathbf{j}(a_zb_x - a_xb_z) + \mathbf{k}(a_xb_y - a_yb_x) \tag{8-11}$$

A convenient way to remember **A** × **B** is to express it in the form of a determinant (see Chapter 9):

$$\mathbf{A} \times \mathbf{B} = \begin{vmatrix} \mathbf{i} & \mathbf{j} & \mathbf{k} \\ a_x & a_y & a_z \\ b_x & b_y & b_z \end{vmatrix}$$

Example

Determine the vector product between the vectors **A**(4, 3, 2) and **B**(−1, 2, −3).

Solution.

$$\mathbf{A} = \mathbf{i}(4) + \mathbf{j}(3) + \mathbf{k}(2)$$

$$\mathbf{B} = \mathbf{i}(-1) + \mathbf{j}(2) + \mathbf{k}(-3)$$

$$\mathbf{A} \times \mathbf{B} = \begin{vmatrix} \mathbf{i} & \mathbf{j} & \mathbf{k} \\ 4 & 3 & 2 \\ -1 & 2 & -3 \end{vmatrix}$$

$$\begin{aligned} \mathbf{A} \times \mathbf{B} &= \mathbf{i}[3(-3) - 2(2)] - \mathbf{j}[4(-3) - (-1)(2)] + \mathbf{k}[4(2) - (-1)(3)] \\ &= \mathbf{i}(-13) + \mathbf{j}(10) + \mathbf{k}(11) \end{aligned}$$

8-4 APPLICATIONS

In this section we shall consider examples that demonstrate the application of vector analysis to physicochemical systems. The subject of vector operators is discussed in Chapter 10.

1. The interaction between the magnetic moment of a nucleus, $\boldsymbol{\mu}$, and a magnetic field, $\mathbf{H}$, is given by the equation

$$E = -\boldsymbol{\mu} \cdot \mathbf{H}$$

Let us determine the components of energy associated with the interaction along the x-, y-, and z-axes. Since

$$\boldsymbol{\mu} = \mathbf{i}\mu_x + \mathbf{j}\mu_y + \mathbf{k}\mu_z$$

$$\mathbf{H} = \mathbf{i}H_x + \mathbf{j}H_y + \mathbf{k}H_z$$

then

$$\boldsymbol{\mu} \cdot \mathbf{H} = \mu_x H_x + \mu_y H_y + \mu_z H_z$$

Also, since

$$E = E_x + E_y + E_z$$

we can write

$$E_x = -\mu_x H_x, \quad E_y = -\mu_y H_y, \quad \text{and} \quad E_z = -\mu_z H_z$$

Note that although both $\boldsymbol{\mu}$ and $\mathbf{H}$ are vectors, energy is a scalar.

2. The torque exerted on a nucleus having a magnetic moment $\boldsymbol{\mu}$ in a magnetic field $\mathbf{H}$ is

$$\mathbf{T} = -\boldsymbol{\mu} \times \mathbf{H}$$

Torque, however, is the rate of change of angular momentum, $d\mathbf{L}/dt$, where

$$\mathbf{L} = \mathbf{i}L_x + \mathbf{j}L_y + \mathbf{k}L_z$$

and

$$\frac{d\mathbf{L}}{dt} = \mathbf{i}\frac{dL_x}{dt} + \mathbf{j}\frac{dL_y}{dt} + \mathbf{k}\frac{dL_z}{dt}$$

Let us determine the torque on the nucleus along the x-, y-, and z-axes. Since

$$\frac{d\mathbf{L}}{dt} = -\boldsymbol{\mu} \times \mathbf{H}$$

$$\boldsymbol{\mu} \times \mathbf{H} = \begin{vmatrix} \mathbf{i} & \mathbf{j} & \mathbf{k} \\ \mu_x & \mu_y & \mu_z \\ H_x & H_y & H_z \end{vmatrix}$$

$$\begin{aligned} \boldsymbol{\mu} \times \mathbf{H} &= \mathbf{i}[\mu_y H_z - \mu_z H_y] - \mathbf{j}[\mu_x H_z - \mu_z H_x] + \mathbf{k}[\mu_x H_y - \mu_y H_x] \\ &= \mathbf{i}[\mu_y H_z - \mu_z H_y] + \mathbf{j}[\mu_z H_x - \mu_x H_z] + \mathbf{k}[\mu_x H_y - \mu_y H_x] \\ -\boldsymbol{\mu} \times \mathbf{H} &= \mathbf{i}[\mu_z H_y - \mu_y H_z] + \mathbf{j}[\mu_x H_z - \mu_z H_x] + \mathbf{k}[\mu_y H_x - \mu_x H_y] \end{aligned}$$

$$\frac{dL_x}{dt} = \mu_z H_y - \mu_y H_z, \quad \frac{dL_y}{dt} = \mu_x H_z - \mu_z H_x, \quad \text{and} \quad \frac{dL_z}{dt} = \mu_y H_x - \mu_x H_y$$

SUGGESTED READING

1. Bradley, Gerald L., and Smith, Karl J., *Calculus*, Prentice-Hall, Inc., Upper Saddle River, NJ, 1995.
2. Varberg, Dale, and Purcell, Edwin J., *Calculus*, 7th ed., Prentice-Hall, Inc., Upper Saddle River, NJ, 1997.
3. Washington, Allyn J., *Basic Technical Mathematics*, 6th ed., Addison-Wesley Publishing Co., Boston, 1995.

PROBLEMS

1. Determine the magnitude and direction of the following vectors:

(a) $\mathbf{A}(1, 3)$
(b) $\mathbf{A}(2, 2)$
(c) $\mathbf{A}(3, -4)$
(d) $\mathbf{A}(-2, 0)$
(e) $\mathbf{A}(-1, -6)$
(f) $\mathbf{A}(1, 1, 3)$
(g) $\mathbf{A}(2, 3, 4)$
(h) $\mathbf{A}(-1, 2, -1)$
(i) $\mathbf{A}(-1, -1, -3)$
(j) $\mathbf{A}(1, 0, -1)$

2. Determine the magnitude and direction of the following sums:

(a) $\mathbf{A}(1, 3) + \mathbf{B}(3, 1)$
(b) $\mathbf{A}(-1, 2) + \mathbf{B}(2, 2)$
(c) $\mathbf{A}(3, -1) + \mathbf{B}(0, 4)$
(d) $\mathbf{A}(1, 1, 1) + \mathbf{B}(2, 3, 4)$
(e) $\mathbf{A}(-2, 3, 4) + \mathbf{B}(-1, -4, -6)$
(f) $\mathbf{A}(2, 0, 3) + \mathbf{B}(-3, 6, -9)$

3. Find the following scalar products:

(a) $\mathbf{A}(1, 3) \cdot \mathbf{B}(3, 1)$
(b) $\mathbf{A}(-1, 2) \cdot \mathbf{B}(2, 2)$
(c) $\mathbf{A}(3, -1) \cdot \mathbf{B}(0, 4)$
(d) $\mathbf{A}(1, 1, 1) \cdot \mathbf{B}(2, 3, 4)$
(e) $\mathbf{A}(-2, 3, 4) \cdot \mathbf{B}(-1, -4, -6)$
(f) $\mathbf{A}(2, 0, 3) \cdot \mathbf{B}(-3, 6, -9)$

4. Find the magnitude and direction of the following vector products:

(a) $\mathbf{A}(1, 3) \times \mathbf{B}(3, 1)$
(b) $\mathbf{A}(-1, 2) \times \mathbf{B}(2, 2)$
(c) $\mathbf{A}(3, -1) \times \mathbf{B}(0, 4)$
(d) $\mathbf{A}(1, 1, 1) \times \mathbf{B}(2, 3, 4)$
(e) $\mathbf{A}(-2, 3, 4) \times \mathbf{B}(-1, -4, -6)$
(f) $\mathbf{A}(2, 0, 3) \times \mathbf{B}(-3, 6, -9)$

5. Show that

$$\mathbf{A} + (\mathbf{B} + \mathbf{C}) = (\mathbf{A} + \mathbf{B}) + \mathbf{C}$$

6. Show that scalar multiplication is commutative and vector multiplication is not. That is,

$$\mathbf{A} \cdot \mathbf{B} = \mathbf{B} \cdot \mathbf{A} \quad \text{but} \quad \mathbf{A} \times \mathbf{B} \neq \mathbf{B} \times \mathbf{A}$$

7. Show that

$$\mathbf{A} \cdot \mathbf{A} = |A|^2$$

8. Angular momentum is given by the equation $\mathbf{L} = \mathbf{r} \times \mathbf{p}$, where $\mathbf{r} = \mathbf{i}x + \mathbf{j}y + \mathbf{k}z$ is the radius of curvature and $\mathbf{p} = \mathbf{i}p_x + \mathbf{j}p_y + \mathbf{k}p_z$ is the linear momentum. Assuming that

$$\mathbf{L} = \mathbf{i}L_x + \mathbf{j}L_y + \mathbf{k}L_z$$

find the components of angular momentum in the x-, y-, and z-directions.

9. Show that the vectors $\mathbf{A} = \frac{1}{2}\mathbf{q}_1 + \frac{1}{2}\mathbf{q}_2 + \frac{1}{2}\mathbf{q}_3 + \frac{1}{2}\mathbf{q}_4$ and $\mathbf{B} = \frac{1}{2}\mathbf{q}_1 - \frac{1}{2}\mathbf{q}_2 + \frac{1}{2}\mathbf{q}_3 - \frac{1}{2}\mathbf{q}_4$, where $\mathbf{q}_1$, $\mathbf{q}_2$, $\mathbf{q}_3$, and $\mathbf{q}_4$ are unit vectors, are orthogonal.

$$\begin{pmatrix} a_{11} & a_{12} & a_{13} & \cdots & a_{1n} \\ a_{21} & a_{22} & a_{23} & \cdots & a_{2n} \\ a_{31} & a_{32} & a_{33} & \cdots & a_{3n} \\ \vdots & \vdots & \vdots & & \vdots \\ a_{m1} & a_{m2} & a_{m3} & \cdots & a_{mn} \end{pmatrix}$$

9 Matrices and Determinants

9-1 INTRODUCTION

In certain areas of physical chemistry, it is convenient to utilize a two-dimensional array of numbers called a *matrix*. Matrices may be either square, containing an equal number of horizontal and vertical lines, or they may be rectangular. The horizontal lines of the matrix are called *rows*, while the vertical lines are called *columns*. A matrix with m rows and n columns is represented by the expression

$$\mathbf{A} = \begin{pmatrix} a_{11} & a_{12} & a_{13} & \cdots & a_{1n} \\ a_{21} & a_{22} & a_{23} & \cdots & a_{2n} \\ \vdots & \vdots & \vdots & & \vdots \\ a_{m1} & a_{m2} & a_{m3} & \cdots & a_{mn} \end{pmatrix}$$

where $a_{ij} = a_{11}, a_{12}, a_{13}, \ldots$ are known as the *elements* of the matrix. Such a matrix is called either a matrix of order (m, n) or an $m \times n$ matrix. The simplest forms of matrices are the row matrix $\mathbf{B}$ and the column matrix $\mathbf{C}$:

$$\mathbf{B} = (b_1 \quad b_2 \quad b_3 \quad \cdots \quad b_n) \quad \text{and} \quad \mathbf{C} = \begin{pmatrix} c_1 \\ c_2 \\ \vdots \\ c_n \end{pmatrix}$$

Matrices have some very useful properties and an algebra all their own. However, before going into these, let us first concentrate on one specific type of matrix, the square matrix.

9-2 SQUARE MATRICES AND DETERMINANTS

As mentioned, a square matrix is one having an equal number of rows and columns. The number of rows or columns in a square matrix is called the *order* of the matrix. Hence, a third-order matrix is one having three rows and three columns.

Associated with every square matrix is a real number called the *determinant* of the matrix. The determinant of a second-order matrix is defined as

$$D = \begin{vmatrix} a & b \\ c & d \end{vmatrix} = ad - bc \tag{9-1}$$

where a, b, c, and d are the elements of the matrix. It is important to note that the matrix itself has no numerical value. Only the determinant of the matrix can be assigned a specific value. To illustrate the use of Equation (9-1), consider the following examples:

Examples

(a) Evaluate the determinant $\begin{vmatrix} 3 & 1 \\ 4 & 2 \end{vmatrix}$.

Solution:

$$\begin{vmatrix} 3 & 1 \\ 4 & 2 \end{vmatrix} = 3(2) - 4(1) = 2$$

(b) Evaluate the determinant $\begin{vmatrix} 5 & 6 \\ -1 & -4 \end{vmatrix}$.

Solution:

$$\begin{vmatrix} 5 & 6 \\ -1 & -4 \end{vmatrix} = 5(-4) - (-1)(6) = -14$$

(c) Evaluate the determinant $\begin{vmatrix} \sin\theta & \cos\theta \\ -\cos\theta & \sin\theta \end{vmatrix}$.

Solution:

$$\begin{vmatrix} \sin\theta & \cos\theta \\ -\cos\theta & \sin\theta \end{vmatrix} = \sin^2\theta + \cos^2\theta = 1$$

In order to evaluate determinants of orders higher than order 2, we use the method of cofactors, described as follows. Consider the determinant

$$D = \begin{vmatrix} a_{11} & a_{12} & a_{13} \\ a_{21} & a_{22} & a_{23} \\ a_{31} & a_{32} & a_{33} \end{vmatrix}$$

The cofactor of element a_{ij} is equal to $(-1)^{i+j}$ multiplied by the determinant that is formed by eliminating the *ith* row and the *jth* column from the original determinant. The determinant then is expanded by summing together the elements multiplied by their cofactors for any row. Hence, using the first row, for example, we can write

$$D = \begin{vmatrix} a_{11} & a_{12} & a_{13} \\ a_{21} & a_{22} & a_{23} \\ a_{31} & a_{32} & a_{33} \end{vmatrix} = a_{11}\begin{vmatrix} a_{22} & a_{23} \\ a_{32} & a_{33} \end{vmatrix} - a_{12}\begin{vmatrix} a_{21} & a_{23} \\ a_{31} & a_{33} \end{vmatrix} + a_{13}\begin{vmatrix} a_{21} & a_{22} \\ a_{31} & a_{32} \end{vmatrix}$$

Similar expressions are possible using the elements of the second row or third row as cofactors.

If each new determinant that results in the expansion by cofactors is one of higher order than 2, then the process is repeated for each of these determinants until the resulting determinants are of order 2. Since, in this example, the resulting determinants are of order 2, we have

$$D = a_{11}(a_{22}a_{33} - a_{32}a_{23}) - a_{12}(a_{21}a_{33} - a_{31}a_{23}) + a_{13}(a_{21}a_{32} - a_{31}a_{22})$$

Example

Evaluate the determinant $\begin{vmatrix} 1 & 4 & 3 & 2 \\ 6 & 1 & 1 & 3 \\ -1 & 4 & 5 & -6 \\ 2 & 1 & 2 & -3 \end{vmatrix}$.

Solution. Expanding by cofactors, we have

$$\begin{vmatrix} 1 & 4 & 3 & 2 \\ 6 & 1 & 1 & 3 \\ -1 & 4 & 5 & -6 \\ 2 & 1 & 2 & -3 \end{vmatrix} = (1)\begin{vmatrix} 1 & 1 & 3 \\ 4 & 5 & -6 \\ 1 & 2 & -3 \end{vmatrix} - (4)\begin{vmatrix} 6 & 1 & 3 \\ -1 & 5 & -6 \\ 2 & 2 & -3 \end{vmatrix}$$

$$+ (3)\begin{vmatrix} 6 & 1 & 3 \\ -1 & 4 & -6 \\ 2 & 1 & -3 \end{vmatrix} - (2)\begin{vmatrix} 6 & 1 & 1 \\ -1 & 4 & 5 \\ 2 & 1 & 2 \end{vmatrix}$$

$$= 1\begin{vmatrix} 5 & -6 \\ 2 & -3 \end{vmatrix} - 1\begin{vmatrix} 4 & -6 \\ 1 & -3 \end{vmatrix} + 3\begin{vmatrix} 4 & 5 \\ 1 & 2 \end{vmatrix} - 24\begin{vmatrix} 5 & -6 \\ 2 & -3 \end{vmatrix}$$

$$+ 4\begin{vmatrix} -1 & -6 \\ 2 & -3 \end{vmatrix} - 12\begin{vmatrix} -1 & 5 \\ 2 & 2 \end{vmatrix} + 18\begin{vmatrix} 4 & -6 \\ 1 & -3 \end{vmatrix}$$

$$- 3\begin{vmatrix} -1 & -6 \\ 2 & -3 \end{vmatrix} + 9\begin{vmatrix} -1 & 4 \\ 2 & 1 \end{vmatrix} - 12\begin{vmatrix} 4 & 5 \\ 1 & 2 \end{vmatrix} + 2\begin{vmatrix} -1 & 5 \\ 2 & 2 \end{vmatrix}$$

$$- 2\begin{vmatrix} -1 & 4 \\ 2 & 1 \end{vmatrix}$$

$$\begin{aligned} &= 1(-3) - 1(-6) + 3(3) - 24(-3) + 4(15) - 12(-12) \\ &\quad + 18(-6) - 3(15) + 9(-9) - 12(3) + 2(-12) \\ &\quad - 2(-9) = 12 \end{aligned}$$

9-3 MATRIX ALGEBRA

Let us turn now to several rules that govern the properties of matrices and their determinants. These rules are presented without proof.

1. *If the corresponding rows and columns of a square matrix are interchanged, the determinant of the matrix remains unchanged.*

$$\begin{vmatrix} a_{11} & a_{12} & a_{13} \\ a_{21} & a_{22} & a_{23} \\ a_{31} & a_{32} & a_{33} \end{vmatrix} = \begin{vmatrix} a_{11} & a_{21} & a_{31} \\ a_{12} & a_{22} & a_{32} \\ a_{13} & a_{23} & a_{33} \end{vmatrix}$$

2. *If any two rows or columns of a determinant are interchanged, the sign of the determinant changes.*

$$\begin{vmatrix} a_{11} & a_{12} & a_{13} \\ a_{21} & a_{22} & a_{23} \\ a_{31} & a_{32} & a_{33} \end{vmatrix} = - \begin{vmatrix} a_{11} & a_{13} & a_{12} \\ a_{21} & a_{23} & a_{22} \\ a_{31} & a_{33} & a_{32} \end{vmatrix}$$

3. *If any two rows or columns of a square matrix are identical, its determinant is zero.*

$$\begin{vmatrix} a_{11} & a_{11} & a_{13} \\ a_{21} & a_{21} & a_{23} \\ a_{31} & a_{31} & a_{33} \end{vmatrix} = 0$$

4. *If each element in any row or column in a determinant is multiplied by the same number k, the value of the determinant is multiplied by k.*

$$\begin{vmatrix} a_{11} & a_{12} & a_{13} \\ ka_{21} & ka_{22} & ka_{23} \\ a_{31} & a_{32} & a_{33} \end{vmatrix} = k \begin{vmatrix} a_{11} & a_{12} & a_{13} \\ a_{21} & a_{22} & a_{23} \\ a_{31} & a_{32} & a_{33} \end{vmatrix}$$

5. *If each element in any row or column in a determinant is multiplied by the same number k and the product is added to a corresponding element in another column, the value of the determinant remains unchanged.*

$$\begin{vmatrix} a_{11} + ka_{12} & a_{12} & a_{13} \\ a_{21} + ka_{22} & a_{22} & a_{23} \\ a_{31} + ka_{32} & a_{32} & a_{33} \end{vmatrix} = \begin{vmatrix} a_{11} & a_{12} & a_{13} \\ a_{21} & a_{22} & a_{23} \\ a_{31} & a_{32} & a_{33} \end{vmatrix}$$

6. *Two matrices are added by the addition of their elements.*

$$\begin{pmatrix} a_{11} & a_{12} \\ a_{21} & a_{22} \end{pmatrix} + \begin{pmatrix} b_{11} & b_{12} \\ b_{21} & b_{22} \end{pmatrix} = \begin{pmatrix} a_{11} + b_{11} & a_{12} + b_{12} \\ a_{21} + b_{21} & a_{22} + b_{22} \end{pmatrix}$$

7. *Two matrices are multiplied according to the following method:*

$$\begin{pmatrix} a_{11} & a_{12} & a_{13} \\ a_{21} & a_{22} & a_{23} \\ a_{31} & a_{32} & a_{33} \end{pmatrix} \begin{pmatrix} b_{11} & b_{12} & b_{13} \\ b_{21} & b_{22} & b_{23} \\ b_{31} & b_{32} & b_{33} \end{pmatrix} =$$

$$\begin{pmatrix} a_{11}\, b_{11} + a_{12}\, b_{21} + a_{13}\, b_{31} & a_{11}\, b_{12} + a_{12}\, b_{22} + a_{13}\, b_{32} & a_{11}\, b_{13} + a_{12}\, b_{23} + a_{13}\, b_{33} \\ a_{21}\, b_{11} + a_{22}\, b_{21} + a_{23}\, b_{31} & a_{21}\, b_{12} + a_{22}\, b_{22} + a_{23}\, b_{32} & a_{21}\, b_{13} + a_{22}\, b_{23} + a_{23}\, b_{33} \\ a_{31}\, b_{11} + a_{32}\, b_{21} + a_{33}\, b_{31} & a_{31}\, b_{12} + a_{32}\, b_{22} + a_{33}\, b_{32} & a_{31}\, b_{13} + a_{32}\, b_{23} + a_{33}\, b_{33} \end{pmatrix}$$

We find from experience that matrix multiplication is defined only if the number of columns of the first matrix equals the number of rows of the second matrix. Moreover, if **A** is an $m \times r$ matrix with elements a_{ij} and **B** is an $r \times n$ matrix with elements b_{ij}, then the resulting product $\mathbf{C} = \mathbf{AB}$ will be an $m \times n$ matrix with elements

$$c_{ij} = \sum_k a_{ik} b_{kj}$$

It is important to note that, of the seven rules given above, this seventh rule should be stressed as the most basic matrix property, since as a natural consequence of this method of multiplication, we find that matrix multiplication is not necessarily commutative. That is, **AB** does not necessarily equal **BA**. This basic property of matrices ultimately led Werner Heisenberg and Max Born to what we now refer to as the Heisenberg uncertainty principle.

Examples

(a) $$\begin{pmatrix} 1 & 2 \\ 3 & 4 \end{pmatrix} \begin{pmatrix} 5 & 6 \\ 7 & 8 \end{pmatrix} = \begin{pmatrix} 1(5) + 2(7) & 1(6) + 2(8) \\ 3(5) + 4(7) & 3(6) + 4(8) \end{pmatrix} = \begin{pmatrix} 19 & 22 \\ 43 & 50 \end{pmatrix}$$

(b) $$\begin{pmatrix} a_{11} & a_{12} \\ a_{21} & a_{22} \end{pmatrix} \begin{pmatrix} c_1 \\ c_2 \end{pmatrix} = \begin{pmatrix} a_{11} & a_{12} \\ a_{21} & a_{22} \end{pmatrix} \begin{pmatrix} c_1 & 0 \\ c_2 & 0 \end{pmatrix} = \begin{pmatrix} a_{11}\, c_1 + a_{12}\, c_2 \\ a_{21}\, c_1 + a_{22}\, c_2 \end{pmatrix}$$

(c) $$\begin{pmatrix} a_1 & a_2 \end{pmatrix} \begin{pmatrix} c_1 \\ c_2 \end{pmatrix} = \begin{pmatrix} a_1 & a_2 \\ 0 & 0 \end{pmatrix} \begin{pmatrix} c_1 & 0 \\ c_2 & 0 \end{pmatrix} = \begin{pmatrix} a_1 c_1 + a_2 c_2 \end{pmatrix}$$

A special type of matrix multiplication occurs when the matrices have nonzero elements in blocks along the diagonal. For example, consider the multiplication of the matrices

$$\left(\begin{array}{cc|c|cc} 1 & 2 & 0 & 0 & 0 \\ 3 & 4 & 0 & 0 & 0 \\ \hline 0 & 0 & 8 & 0 & 0 \\ \hline 0 & 0 & 0 & 1 & 4 \\ 0 & 0 & 0 & -5 & 6 \end{array}\right) \cdot \left(\begin{array}{cc|c|cc} -3 & -3 & 0 & 0 & 0 \\ 4 & 0 & 0 & 0 & 0 \\ \hline 0 & 0 & 2 & 0 & 0 \\ \hline 0 & 0 & 0 & 4 & -1 \\ 0 & 0 & 0 & -1 & 3 \end{array}\right)$$

In this special case each set of blocks can be multiplied separately:

$$\begin{pmatrix} 1 & 2 \\ 3 & 4 \end{pmatrix} \begin{pmatrix} -3 & -3 \\ 4 & 0 \end{pmatrix} = \begin{pmatrix} 5 & -3 \\ 7 & -9 \end{pmatrix}$$

$$(8)(2) = (16)$$

$$\begin{pmatrix} 1 & 4 \\ -5 & 6 \end{pmatrix} \begin{pmatrix} 4 & -1 \\ -1 & 3 \end{pmatrix} = \begin{pmatrix} 0 & 11 \\ -26 & 23 \end{pmatrix}$$

giving the answer

$$\begin{pmatrix} 5 & -3 & 0 & 0 & 0 \\ 7 & -9 & 0 & 0 & 0 \\ 0 & 0 & 16 & 0 & 0 \\ 0 & 0 & 0 & 0 & 11 \\ 0 & 0 & 0 & -26 & 23 \end{pmatrix}$$

A matrix having nonzero values for elements along its diagonal and zero for all other elements is known as a *diagonal matrix*. When the nonzero values along the diagonal are equal to unity, the matrix is called a *unit matrix*. Hence, a 3×3 unit matrix

$$\mathbf{1} = \begin{pmatrix} 1 & 0 & 0 \\ 0 & 1 & 0 \\ 0 & 0 & 1 \end{pmatrix}$$

We can show by matrix multiplication that if **A** is any matrix, then

$$\mathbf{1}\ \mathbf{A} = \mathbf{A}$$

If **A** is a square matrix and the determinant of **A** does not equal zero, then **A** is said to be nonsingular.[1] If **A** is a nonsingular matrix, then there exists an inverse matrix $\mathbf{A}^{-1}$ such that

$$\mathbf{A}^{-1}\mathbf{A} = \mathbf{A}\mathbf{A}^{-1} = \mathbf{1}$$

The inverse of a matrix can be found by solving the equation $\mathbf{A}^{-1}\mathbf{A} = \mathbf{1}$. Examples of a nonsingular matrix and its inverse are

$$\begin{pmatrix} 0 & -1 \\ 1 & 0 \end{pmatrix} \quad \text{and} \quad \begin{pmatrix} 0 & 1 \\ -1 & 0 \end{pmatrix}$$

[1]A singular matrix is defined as one whose determinant is equal to zero.

9-4 SOLUTIONS OF SYSTEMS OF LINEAR EQUATIONS

Consider the following set of linear equations

$$\begin{array}{l} a_{11}x_1 + a_{12}x_2 + a_{13}x_3 + \cdots + a_{1n}x_n = c_1 \\ a_{21}x_1 + a_{22}x_2 + a_{23}x_3 + \cdots + a_{2n}x_n = c_2 \\ \cdot \quad \cdot \quad \cdot \quad \cdot \quad \cdot \\ \cdot \quad \cdot \quad \cdot \quad \cdot \quad \cdot \\ \cdot \quad \cdot \quad \cdot \quad \cdot \quad \cdot \\ a_{n1}x_1 + a_{n2}x_2 + a_{n3}x_3 + \cdots + a_{nn}x_n = c_n \end{array} \tag{9-2}$$

We can easily show that the set of Equations (9-2) can be represented by the product of two matrices

$$\begin{pmatrix} a_{11} & a_{12} & a_{13} & \cdots & a_{1n} \\ a_{21} & a_{22} & a_{23} & \cdots & a_{2n} \\ \cdot & \cdot & \cdot & & \cdot \\ \cdot & \cdot & \cdot & & \cdot \\ \cdot & \cdot & \cdot & & \cdot \\ a_{n1} & a_{n2} & a_{n3} & \cdots & a_{nn} \end{pmatrix} \begin{pmatrix} x_1 \\ x_2 \\ \cdot \\ \cdot \\ \cdot \\ x_n \end{pmatrix} = \begin{pmatrix} c_1 \\ c_2 \\ \cdot \\ \cdot \\ \cdot \\ c_n \end{pmatrix} \tag{9-3}$$

The determinant of the coefficients is

$$D = \begin{vmatrix} a_{11} & a_{12} & a_{13} & \cdots & a_{1n} \\ a_{21} & a_{22} & a_{23} & \cdots & a_{2n} \\ \cdot & \cdot & \cdot & & \cdot \\ \cdot & \cdot & \cdot & & \cdot \\ \cdot & \cdot & \cdot & & \cdot \\ a_{n1} & a_{n2} & a_{n3} & \cdots & a_{nn} \end{vmatrix}$$

From Rule 4 in the previous section, we can write

$$Dx_1 = \begin{vmatrix} a_{11}x_1 & a_{12} & a_{13} & \cdots & a_{1n} \\ a_{21}x_1 & a_{22} & a_{23} & \cdots & a_{2n} \\ \cdot & \cdot & \cdot & & \cdot \\ \cdot & \cdot & \cdot & & \cdot \\ \cdot & \cdot & \cdot & & \cdot \\ a_{n1}x_1 & a_{n2} & a_{n3} & \cdots & a_{nn} \end{vmatrix}$$

If we now multiply each element in column 2 by x_2, each element in column 3 by x_3, and so on, and add these to column 1 in determinant Dx_1, then by Rule 5 in the previous section Dx_1 remains unchanged.

$$Dx_1 = \begin{vmatrix} a_{11}x_1 + a_{12}x_2 + \cdots + a_{1n}x_n & a_{12} & a_{13} & \cdots & a_{1n} \\ a_{21}x_1 + a_{22}x_2 + \cdots + a_{2n}x_n & a_{22} & a_{23} & \cdots & a_{2n} \\ \cdot & \cdot & \cdot & & \cdot \\ \cdot & \cdot & \cdot & & \cdot \\ \cdot & \cdot & \cdot & & \cdot \\ a_{n1}x_1 + a_{n2}x_2 + \cdots + a_{nn}x_n & a_{n2} & a_{n3} & \cdots & a_{nn} \end{vmatrix}$$

Substituting for column 1 the Equations (9-2), we have

$$Dx_1 = \begin{vmatrix} c_1 & a_{12} & a_{13} & \cdots & a_{1n} \\ c_2 & a_{22} & a_{23} & \cdots & a_{2n} \\ \cdot & \cdot & \cdot & & \cdot \\ \cdot & \cdot & \cdot & & \cdot \\ \cdot & \cdot & \cdot & & \cdot \\ c_n & a_{n2} & a_{n3} & \cdots & a_{nn} \end{vmatrix} = D_1$$

By the same argument used for column 1, we can write $Dx_2 = D_2$, $Dx_3 = D_3$, and so on, where D_i is determinant D in which the elements in column i have been replaced by $c_1, c_2, c_3, c_4, \ldots$, etc. Hence, if $D \neq 0$, we have the solutions to the set of equations

$$x_1 = \frac{D_1}{D}, \quad x_2 = \frac{D_2}{D}, \quad x_3 = \frac{D_3}{D}, \quad \text{and so on.}$$

This method of solving sets of linear equations is known as solution by *Cramer's rule*.

Example

Solve the following equations using Cramer's rule:

$$\begin{aligned} x + y + z &= 2 \\ 3x + y - 2z &= -5 \\ 2x - y - 3z &= -5 \end{aligned}$$

Solution. We first set up the determinants:

$$D = \begin{vmatrix} 1 & 1 & 1 \\ 3 & 1 & -2 \\ 2 & -1 & -3 \end{vmatrix} = -5 \qquad D_1 = \begin{vmatrix} 2 & 1 & 1 \\ -5 & 1 & -2 \\ -5 & -1 & -3 \end{vmatrix} = -5$$

$$D_2 = \begin{vmatrix} 1 & 2 & 1 \\ 3 & -5 & -2 \\ 2 & -5 & -3 \end{vmatrix} = 10 \qquad D_3 = \begin{vmatrix} 1 & 1 & 2 \\ 3 & 1 & -5 \\ 2 & -1 & -5 \end{vmatrix} = -15$$

$$x = \frac{-5}{-5} = 1 \qquad y = \frac{10}{-5} = -2 \qquad z = \frac{-15}{-5} = 3$$

Consider, now, a special set of linear equations that are all equal to zero.

$$\begin{array}{ccccccccccc} a_{11}x_1 & + & a_{12}x_2 & + & a_{13}x_3 & + & \cdots & + & a_{1n}x_n & = & 0 \\ a_{21}x_1 & + & a_{22}x_2 & + & a_{23}x_3 & + & \cdots & + & a_{2n}x_n & = & 0 \\ \cdot & & \cdot & & \cdot & & \cdot & & \cdot & & \\ \cdot & & \cdot & & \cdot & & \cdot & & \cdot & & \\ \cdot & & \cdot & & \cdot & & \cdot & & \cdot & & \\ a_{n1}x_1 & + & a_{n2}x_2 & + & a_{n3}x_3 & + & \cdots & + & a_{nn}x_n & = & 0 \end{array} \tag{9-4}$$

Writing these equations in matrix form, we have

$$\begin{pmatrix} a_{11} & a_{12} & a_{13} & \cdots & a_{1n} \\ a_{21} & a_{22} & a_{23} & \cdots & a_{2n} \\ \cdot & \cdot & \cdot & & \cdot \\ \cdot & \cdot & \cdot & & \cdot \\ \cdot & \cdot & \cdot & & \cdot \\ a_{n1} & a_{n2} & a_{n3} & \cdots & a_{nn} \end{pmatrix} \begin{pmatrix} x_1 \\ x_2 \\ \cdot \\ \cdot \\ \cdot \\ x_n \end{pmatrix} = 0$$

or

$$\mathbf{Ax} = 0$$

If we tried to solve this set of equations using Cramer's rule, we would obtain a trivial set of solutions $x_1 = x_2 = x_3 = \cdots = x_n = 0$. If we wish a nontrivial set of solutions, that is, a set of solutions for which **x** is not equal to zero, then **A** must be singular. Recall that a singular matrix is one whose determinant is equal to zero. Let us explore this idea further by looking at the characteristic equation of a matrix.

9-5 CHARACTERISTIC EQUATION OF A MATRIX

Consider a set of linear equations, such as those described in Section 9-4, written in matrix form as

$$\mathbf{Ac} = \lambda\mathbf{c} \tag{9-5}$$

where λ is a set of scalar constants, called the *eigenvalues* of matrix **A**, and **c** is a column matrix, a vector (see Chapter 8) called the *eigenvector*, belonging to matrix **A**. The only effect that matrix **A** has on matrix **c** is to multiply each element of matrix **c** by a constant scalar λ. It is easy to show, then, that **c** and $\lambda\mathbf{c}$ are parallel vectors in space. Thus, the constant scalar factor λ changes the length of the vector, but not its direction.

Equation (9-5) can be written as

$$(\mathbf{A} - \lambda)\mathbf{c} = 0 \tag{9-6}$$

We saw in Section 9-4, however, that in order not to obtain a trivial set of solutions to the linear equations represented by Equation (9-6), that is, $c_1 = c_2 = c_3 = \cdots = 0$, it must be true that the matrix $(\mathbf{A} - \lambda)$ be singular; that is, det $(\mathbf{A} - \lambda) = 0$. Writing the determinant $(\mathbf{A} - \lambda)$ in explicit form, we have

$$\begin{vmatrix} a_{11} - \lambda_1 & a_{12} & \cdots & a_{1n} \\ a_{21} & a_{22} - \lambda_2 & \cdots & a_{2n} \\ \vdots & \vdots & & \vdots \\ a_{n1} & a_{n2} & \cdots & a_{nn} - \lambda_n \end{vmatrix} = 0 \tag{9-7}$$

The determinant shown in Equation (9-7) is called the *secular determinant,* and the linear equations represented by Equation (9-6) are called the *secular equations.* We can solve this $n \times n$ determinant using the method of cofactors, which will yield an nth order polynomial

$$\lambda_i^n + a_1\lambda_i^{n-1} + a_2\lambda_i^{n-2} + \cdots + a_{n-1}\lambda_i + a_n = 0 \tag{9-8}$$

Equation (9-8) is called the *characteristic equation* of matrix **A**. The roots to Equation (9-8) are called the *eigenvalue spectrum* of matrix **A**. For every eigenvalue λ_i, there exists a corresponding eigenvector $\mathbf{c}_i$. Therefore, if there are n eigenvalues, then there is not one eigenvalue equation, but n eigenvalue equations

$$\begin{aligned} \mathbf{A}\mathbf{c}_1 &= \lambda_1\mathbf{c}_1 \\ \mathbf{A}\mathbf{c}_2 &= \lambda_2\mathbf{c}_2 \\ \mathbf{A}\mathbf{c}_3 &= \lambda_3\mathbf{c}_3 \\ &\vdots \\ \mathbf{A}\mathbf{c}_n &= \lambda_n\mathbf{c}_n \end{aligned} \tag{9-9}$$

where each $\mathbf{c}_i$ vector is a column matrix. Equations (9-9) can be represented by a single matrix equation

$$\mathbf{AC} = \mathbf{C\Lambda} \tag{9-10}$$

or

$$\begin{pmatrix} a_{11} & a_{12} & \cdots & a_{1n} \\ a_{21} & a_{22} & \cdots & a_{2n} \\ \vdots & \vdots & & \vdots \\ a_{n1} & a_{n2} & \cdots & a_{nn} \end{pmatrix} \begin{pmatrix} c_{11} & c_{12} & \cdots & c_{1n} \\ c_{21} & c_{22} & \cdots & c_{21} \\ \vdots & \vdots & & \vdots \\ c_{n1} & c_{n2} & \cdots & c_{nn} \end{pmatrix}$$

$$= \begin{pmatrix} c_{11} & c_{12} & \cdots & c_{1n} \\ c_{21} & c_{22} & \cdots & c_{21} \\ \vdots & \vdots & & \vdots \\ c_{n1} & c_{n2} & \cdots & c_{nn} \end{pmatrix} \begin{pmatrix} \lambda_1 & 0 & \cdots & 0 \\ 0 & \lambda_2 & \cdots & 0 \\ \vdots & \vdots & & \vdots \\ 0 & 0 & \cdots & \lambda_n \end{pmatrix}$$

If det $\mathbf{C} \neq 0$, that is, if $\mathbf{C}$ is nonsingular, then there exists an inverse $\mathbf{C}^{-1}$, and multiplying through Equation (9-10) by $\mathbf{C}^{-1}$ gives

$$\mathbf{C}^{-1}\mathbf{AC} = \mathbf{C}^{-1}\mathbf{C}\boldsymbol{\Lambda} = \boldsymbol{\Lambda} \tag{9-11}$$

We see, then, that we can diagonalize matrix $\mathbf{A}$ by compounding the eigenvectors into a matrix. Finding the eigenvectors that diagonalize a matrix to give its eigenvalues is one of the major types of problems found in quantum mechanics. More will be said about eigenvalues and their relationship to operators in Chapter 10.

Example

A problem in simple Hückel molecular orbital theory requires putting in diagonal form the matrix describing the set of secular equations, an example of which is

$$\begin{pmatrix} \alpha - E & \beta & 0 \\ \beta & \alpha - E & \beta \\ 0 & \beta & \alpha - E \end{pmatrix} \begin{pmatrix} c_1 \\ c_2 \\ c_3 \end{pmatrix} = 0$$

in order to determine the eigenvalues E. Here α and β are integrals, called the *Coulomb integral* and *resonance integral*, respectively. This problem differs somewhat from the general description given above in that, in this case, all the values of E may not necessarily be distinct. When the eigenvalues are not all distinct, it may not be possible to put the matrix in true diagonal form. Moreover, in a Hückel molecular orbital problem, all of the secular equations are not independent. This presents a special problem with which we must deal. In the example shown here, though, all the eigenvalues are distinct. In order not to obtain a trivial solution to the secular equations, that is, $c_1 = c_2 = c_3 = 0$, it must be true that the square matrix be singular. That is,

$$\begin{vmatrix} \alpha - E & \beta & 0 \\ \beta & \alpha - E & \beta \\ 0 & \beta & \alpha - E \end{vmatrix} = 0$$

This determinant is most easily solved if we divide through the determinant by β^3 and let $(\alpha - E)/\beta = x$.

$$\beta^3 \begin{vmatrix} \dfrac{\alpha - E}{\beta} & 1 & 0 \\ 1 & \dfrac{\alpha - E}{\beta} & 1 \\ 0 & 1 & \dfrac{\alpha - E}{\beta} \end{vmatrix} = 0 = \begin{vmatrix} x & 1 & 0 \\ 1 & x & 1 \\ 0 & 1 & x \end{vmatrix}$$

We find, using the method of cofactors, that the characteristic equation of the matrix is

$$x^3 - 2x = 0$$

which has the roots $x = 0, \pm\sqrt{2}$. Each root corresponds to a specific eigenvalue (energy state in this case): $E_1 = \alpha + \sqrt{2}\beta$, $E_2 = 0$, $E_3 = \alpha - \sqrt{2}\beta$. To find the coefficients associated with each energy state, we substitute each energy, in turn, into the secular equations and solve for the coefficients. For example, substituting $E_1 = \alpha + \sqrt{2}\beta$ into the secular equations

$$\begin{aligned}(\alpha - E)\,c_1 + \beta c_2 &= 0\\ \beta c_1 + (\alpha - E)\,c_2 + \beta c_3 &= 0\\ \beta c_2 + (\alpha - E)\,c_3 &= 0\end{aligned}$$

gives $c_1 = (1/\sqrt{2})c_2$ and $c_1 = c_3$. Note that because all three secular equations are not independent, we can determine only the ratio of the coefficients and not their values. If the eigenvectors are normalized, however, then we have another equation connecting the coefficients

$$\sum c_i^2 = 1$$

Therefore, in this example, ${c_1}^2 + {c_2}^2 + {c_3}^2 = 1$. Substituting the values given above into this equation gives $c_1 = c_3 = 1/2$ and $c_2 = \sqrt{2}/2$. Coefficients for the other eigenvalues can be found in a similar manner.

SUGGESTED READING

1. NAGLE, R. KENT, and SAFF, EDWARD B., *Fundamentals of Differential Equations and Boundary Value Problems*, 2nd. ed., Addison-Wesley Publishing Co., Boston, 1996.
2. SULLIVAN, MICHAEL, *College Algebra*, 4th ed., Prentice-Hall, Inc., Upper Saddle River, NJ, 1996.
3. WASHINGTON, ALLYN J., *Basic Technical Mathematics*, 6th ed., Addison-Wesley Publishing Co., Boston, 1995.

PROBLEMS

1. Evaluate the following determinants:

(a) $\begin{vmatrix} 1 & 2 \\ 3 & 4 \end{vmatrix}$ **(b)** $\begin{vmatrix} 6 & 1 \\ -1 & -1 \end{vmatrix}$ **(c)** $\begin{vmatrix} 4 & -3 \\ 0 & -1 \end{vmatrix}$

(d) $\begin{vmatrix} 1 & 0 \\ 0 & 1 \end{vmatrix}$ **(e)** $\begin{vmatrix} x & 1 \\ 1 & x \end{vmatrix}$ **(f)** $\begin{vmatrix} \sec\theta & \tan\theta \\ \tan\theta & \sec\theta \end{vmatrix}$

(g) $\begin{vmatrix} 1 & 2 & 3 \\ 3 & 0 & 1 \\ -1 & 4 & 2 \end{vmatrix}$ **(h)** $\begin{vmatrix} 4 & 2 & -1 \\ -1 & 6 & 3 \\ -1 & 5 & -1 \end{vmatrix}$ **(i)** $\begin{vmatrix} x & 1 & 0 \\ 1 & x & 1 \\ 0 & 1 & x \end{vmatrix}$

(j) $\begin{vmatrix} 4 & 3 & 1 & -1 \\ 6 & 1 & 0 & -3 \\ 1 & 5 & 2 & -2 \\ 8 & 6 & -5 & 0 \end{vmatrix}$ **(k)** $\begin{vmatrix} x & b & 0 & 0 \\ b & x & b & 0 \\ 0 & b & x & b \\ 0 & 0 & b & x \end{vmatrix}$

2. Solve the following determinants for x:

(a) $\begin{vmatrix} x & 1 \\ 1 & x \end{vmatrix} = 0$ **(b)** $\begin{vmatrix} x & -2 \\ 1 & x \end{vmatrix} = 6$ **(c)** $\begin{vmatrix} 2x & 4 \\ 2 & x \end{vmatrix} = 2$

(d) $\begin{vmatrix} x & 1 & 1 \\ 1 & x & 1 \\ 1 & 1 & x \end{vmatrix} = 2$ **(e)** $\begin{vmatrix} x & 1 & 1 & 1 \\ 1 & x & 0 & 0 \\ 1 & 0 & x & 0 \\ 1 & 0 & 0 & x \end{vmatrix} = 0$

3. Add the matrices:

$$\begin{pmatrix} 1 & 1 & 4 & 3 \\ -1 & 0 & 1 & 2 \\ -1 & 2 & 4 & -3 \\ 5 & 6 & 3 & 5 \end{pmatrix} + \begin{pmatrix} 4 & 0 & -4 & 3 \\ 6 & 3 & -7 & 5 \\ -1 & 1 & -1 & 0 \\ -5 & 2 & 6 & 7 \end{pmatrix}$$

4. Perform the following matrix multiplication:

(a) $\begin{pmatrix} 1 & 4 \\ 3 & 2 \end{pmatrix}\begin{pmatrix} 6 & -3 \\ -3 & 1 \end{pmatrix}$ **(b)** $\begin{pmatrix} 1 & 0 \\ 0 & 1 \end{pmatrix}\begin{pmatrix} 4 & -1 \\ 2 & 3 \end{pmatrix}$

(c) $\begin{pmatrix} 3 & 0 & 3 \\ 4 & -1 & -1 \\ 1 & 2 & 5 \end{pmatrix}\begin{pmatrix} 1 & 1 & 1 \\ -2 & 1 & 6 \\ 3 & 4 & 5 \end{pmatrix}$

(d) $\begin{pmatrix} 1 & 4 & 1 \\ 0 & 1 & 2 \\ 2 & 4 & -3 \end{pmatrix}\begin{pmatrix} 0 & -4 & 3 \\ 6 & 3 & -7 \\ 2 & 6 & 7 \end{pmatrix}$ **(e)** $\begin{pmatrix} 1 & 8 & 4 \\ -2 & 3 & 0 \\ 5 & -1 & -1 \end{pmatrix}\begin{pmatrix} x \\ y \\ z \end{pmatrix}$

5. Given the two matrices

$$\mathbf{A} = \begin{pmatrix} 1 & 1 & 4 \\ 2 & -6 & 10 \\ 4 & -1 & -1 \end{pmatrix} \quad \text{and} \quad \mathbf{B} = \begin{pmatrix} 6 & 1 & 0 \\ 4 & 2 & -1 \\ 8 & -4 & 3 \end{pmatrix}$$

show that $\mathbf{A\,B} \neq \mathbf{B\,A}$.

6. Solve the following sets of equations using Cramer's rule:

(a)
$$\begin{aligned} x + y &= 3 \\ 4x - 3y &= 5 \end{aligned}$$

(b)
$$\begin{aligned} x + 2y + 3z &= -5 \\ -x - 3y + z &= -14 \\ 2x + y + z &= 1 \end{aligned}$$

(c)
$$\begin{aligned} x &+2y &- z &+ t &= 2 \\ x &-2y &+ z &-3t &= 6 \\ 2x &+ y &+2z &+ t &= -4 \\ 3x &+3y &+ z &-2t &= 10 \end{aligned}$$

(d)
$$\begin{aligned} x\sin\theta &+y\cos\theta &= x' \\ -x\cos\theta &+y\sin\theta &= y' \end{aligned}$$

7. Show that only a trivial solution is possible for the set of equations:

$$x + y = 0$$

$$x - y = 0$$

8. Show that the matrix

$$\mathbf{E} = \begin{pmatrix} 1 & 0 & 0 \\ 0 & 1 & 0 \\ 0 & 0 & 1 \end{pmatrix}$$

will transform the vector $\begin{pmatrix} x \\ y \\ z \end{pmatrix}$ into itself.

9. Show that the matrix

$$\mathbf{C}_2 = \begin{pmatrix} -1 & 0 & 0 \\ 0 & -1 & 0 \\ 0 & 0 & 1 \end{pmatrix}$$

will transform the vector $\begin{pmatrix} x \\ y \\ z \end{pmatrix}$ into $\begin{pmatrix} -x \\ -y \\ z \end{pmatrix}$.

10. Prove that the inverse of the matrix

$$\mathbf{C}_3 = \begin{pmatrix} -\frac{1}{2} & \frac{\sqrt{3}}{2} & 0 \\ -\frac{\sqrt{3}}{2} & -\frac{1}{2} & 0 \\ 0 & 0 & 1 \end{pmatrix} \quad \text{is} \quad \mathbf{C}_3^{-1} = \begin{pmatrix} -\frac{1}{2} & -\frac{\sqrt{3}}{2} & 0 \\ \frac{\sqrt{3}}{2} & -\frac{1}{2} & 0 \\ 0 & 0 & 1 \end{pmatrix}$$

11. Put the following matrix in diagonal form:

$$A = \begin{pmatrix} 1 & \sqrt{6} \\ \sqrt{6} & 2 \end{pmatrix}$$

12. Show that the eigenvectors

$$\mathbf{C} = \begin{pmatrix} \frac{\sqrt{6}}{\sqrt{15}} & \frac{\sqrt{6}}{\sqrt{10}} \\ \frac{3}{\sqrt{15}} & \frac{-2}{\sqrt{10}} \end{pmatrix}$$

will diagonalize matrix **A** in Problem 11. (*Hint*: show by matrix multiplication that $\mathbf{C}^{-1}\mathbf{A}\mathbf{C} = \mathbf{\Lambda}$, where $\mathbf{\Lambda}$ is the diagonal form of **A**.

13. Show that **c** and λ**c**, where λ is a scalar, are parallel vectors in space.

14. Solve the following set of secular equations for E in terms of α and β. Determine the relationship between the coefficients c_1, c_2, and c_3 for each value of E, and using the fact that $\Sigma c_i^2 = 1$ (that is, that the eigenvectors must be normalized), find the values of c_1, c_2, and c_3 for each value of E.

$$\begin{aligned}
(\alpha - E)c_1 + \beta c_2 + \beta c_3 &= 0 \\
\beta c_1 + (\alpha - E)c_2 + \beta c_3 &= 0 \\
\beta c_1 + \beta c_2 + (\alpha - E)c_3 &= 0
\end{aligned}$$

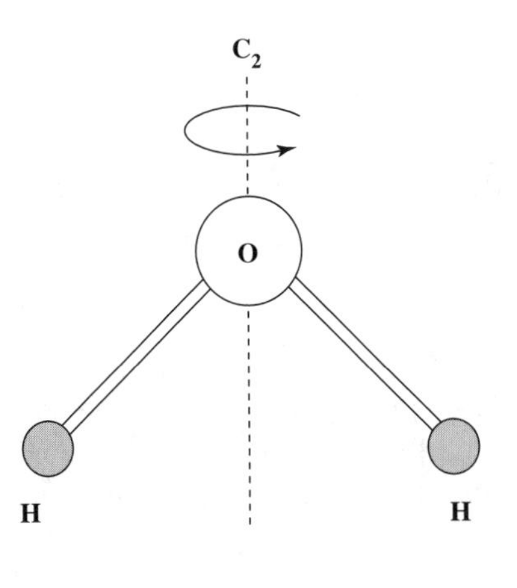

10

Operators

10-1 INTRODUCTION

An *operator* is a symbol that designates a process that will transform one function into another function. For example, the operator $\hat{\mathbf{D}}_x = \partial/\partial x$ applied to the function $f(x)$ designates that the first derivative of $f(x)$ with respect to x should be taken. Some other examples of operators are

$$\boldsymbol{\Delta}: \quad \boldsymbol{\Delta} f(x) = f(x+h) - f(x)$$

$$\boldsymbol{\Sigma}: \quad \boldsymbol{\Sigma} f(x) = f_1(x) + f_2(x) + f_3(x) + \cdots$$

$$\boldsymbol{\Pi}: \quad \boldsymbol{\Pi} f(x) = f_1(x) \cdot f_2(x) \cdot f_3(x) \cdot \cdots$$

If two or more operators are applied simultaneously to a function, then the operator immediately adjacent to the function will operate on the function first, giving a new function; the next adjacent operator will operate on the function next and so on. For example, consider the application of the operators $\hat{\mathbf{D}}_x = \partial/\partial x$ and $\hat{\mathbf{D}}_y = \partial/\partial y$ to the function $f(x, y) = x^3y^2$:

$$\hat{\mathbf{D}}_x \, \hat{\mathbf{D}}_y \, (x^3y^2) = \hat{\mathbf{D}}_x \, (2x^3y) = 6x^2y$$

On the other hand,

$$\hat{\mathbf{D}}_y \, \hat{\mathbf{D}}_x \, (x^3y^2) = \hat{\mathbf{D}}_y \, (3x^2y^2) = 6x^2y$$

When the result of the two operations is independent of the order in which the operators are applied, as in the above example, the operators are said to *commute*.

That is,

$$\hat{\mathbf{D}}_x \, \hat{\mathbf{D}}_y \, f(x, y) = \hat{\mathbf{D}}_y \, \hat{\mathbf{D}}_x \, f(x, y)$$

or

$$[\hat{\mathbf{D}}_x \, \hat{\mathbf{D}}_y - \hat{\mathbf{D}}_y \, \hat{\mathbf{D}}_x] \, f(x, y) = 0 \tag{10-1}$$

Here the brackets [ ] are known as *commutator brackets*, and the expression in the brackets, itself an operator, is called the *commutator of the operators*. It is important to note that the function $f(x, y)$ has not been algebraically factored out of the left-hand side of Equation (10-1). Equation (10-1) is an operator equation and signifies that when the operator $[\hat{\mathbf{D}}_x \, \hat{\mathbf{D}}_y - \hat{\mathbf{D}}_y \, \hat{\mathbf{D}}_x]$ acts on the function $f(x, y)$, it produces the number zero.

In general, then, we can say that two operators $\hat{\mathbf{A}}$ and $\hat{\mathbf{B}}$ commute if, and only if,[1]

$$[\hat{\mathbf{A}} \, \hat{\mathbf{B}} - \hat{\mathbf{B}} \, \hat{\mathbf{A}}] \, f(x) = 0 \tag{10-2}$$

If the commutator does not equal zero, then the order in which the operators are applied must be taken into account.

Certain operators, such as differential operators, may be applied to a function more than once. For example, applying the operator $\hat{\mathbf{D}}_x = \partial/\partial x$ to the function $f(x)$ twice yields

$$\hat{\mathbf{D}}_x \, \hat{\mathbf{D}}_x \, f(x) = \hat{\mathbf{D}}_x^2 \, f(x) = \frac{\partial}{\partial x}\frac{\partial f}{\partial x} = \frac{\partial^2 f}{\partial x^2}$$

which we see is just the second derivative of $f(x)$ with respect to x.

One must be careful, however, in interpreting the square of an operator, especially when the operator contains more than one term. For example, consider the operator

$$\hat{\mathbf{D}}_x = \frac{\partial}{\partial x} = \cos\theta \frac{\partial}{\partial r} - \frac{\sin\theta}{r}\frac{\partial}{\partial \theta}$$

The operator $\hat{\mathbf{D}}_x^2$ implies $\hat{\mathbf{D}}_x$ operating on $\hat{\mathbf{D}}_x$; however, this is not found by simply algebraically squaring $\hat{\mathbf{D}}_x$. On the contrary,

$$\begin{aligned}
\hat{\mathbf{D}}_x^2 &= \left(\cos\theta \frac{\partial}{\partial r} - \frac{\sin\theta}{r}\frac{\partial}{\partial \theta}\right)\left(\cos\theta \frac{\partial}{\partial r} - \frac{\sin\theta}{r}\frac{\partial}{\partial \theta}\right) \\
&= \cos^2\theta \frac{\partial^2}{\partial r^2} - \cos\theta \sin\theta \frac{\partial}{\partial r}\left(\frac{1}{r}\frac{\partial}{\partial \theta}\right) - \frac{\sin\theta}{r}\frac{\partial}{\partial \theta}\left(\cos\theta \frac{\partial}{\partial r}\right) \\
&\quad + \frac{\sin\theta}{r}\frac{\partial}{\partial \theta}\left(\sin\theta \frac{\partial}{\partial \theta}\right)
\end{aligned}$$

[1]Equation (10-2) may be written in an abbreviated form $[\hat{\mathbf{A}}\hat{\mathbf{B}} - \hat{\mathbf{B}}\hat{\mathbf{A}}] = 0$, or even $[\hat{\mathbf{A}}, \hat{\mathbf{B}}] = 0$, provided it is understood what the abbreviated form implies.

Note in the first term that $\cos\theta(\partial/\partial r)$ operates on $\cos\theta(\partial/\partial r)$; however, since $\cos\theta$ is treated as a constant when one differentiates partially with respect to r, it can pass through the operator $\partial/\partial r$ giving

$$\cos^2\theta(\partial/\partial r)(\partial/\partial r)$$

On the other hand, when $\cos\theta(\partial/\partial r)$ operates on $-\dfrac{\sin\theta}{r}\dfrac{\partial}{\partial\theta}$, only the constant $-\sin\theta$ can pass through the operator $\partial/\partial r$. The partial derivative $\partial/\partial\theta$ is not a constant. Hence, the differentiation must be written as

$$-\cos\theta\sin\theta\frac{\partial}{\partial r}\left(\frac{1}{r}\frac{\partial}{\partial\theta}\right)$$

The expression in parentheses must be differentiated as a product. Therefore, we have

$$\begin{aligned}\hat{\mathbf{D}}_x^2 &= \cos^2\theta\frac{\partial^2}{\partial r^2} - \cos\theta\sin\theta\left(\frac{1}{r}\frac{\partial^2}{\partial r\,\partial\theta} - \frac{1}{r^2}\frac{\partial}{\partial\theta}\right)\\ &\quad - \frac{\sin\theta}{r}\left(\cos\theta\frac{\partial^2}{\partial\theta\,\partial r} - \sin\theta\frac{\partial}{\partial r}\right) + \frac{\sin\theta}{r^2}\left(\sin\theta\frac{\partial^2}{\partial\theta^2} + \cos\frac{\partial}{\partial\theta}\right)\\ &= \cos^2\theta\frac{\partial^2}{\partial r^2} - \frac{2\cos\theta\sin\theta}{r}\frac{\partial^2}{\partial r\,\partial\theta} + \frac{2\sin\theta\cos\theta}{r^2}\frac{\partial}{\partial\theta}\\ &\quad + \frac{\sin^2\theta}{r}\frac{\partial}{\partial r} + \frac{\sin^2\theta}{r^2}\frac{\partial^2}{\partial\theta^2}\end{aligned}$$

10-2 VECTOR OPERATORS

One will recall from basic courses in physics that the components of force in the x-, y-, and z-directions, F_x, F_y, and F_z, are related to the potential energy $V(x, y, z)$ by the equations

$$F_x = -\frac{\partial}{\partial x}V(x, y, z)$$

$$F_y = -\frac{\partial}{\partial y}V(x, y, z)$$

$$F_z = -\frac{\partial}{\partial z}V(x, y, z)$$

Since the total force is a vector

$$\mathbf{F} = \mathbf{i}F_x + \mathbf{j}F_y + \mathbf{k}F_y$$

we have

$$\mathbf{F} = -\left(\mathbf{i}\frac{\partial V}{\partial x} + \mathbf{j}\frac{\partial V}{\partial y} + \mathbf{k}\frac{\partial V}{\partial z}\right) \tag{10-3}$$

Equation (10-3), however, can be written in operator form

$$\mathbf{F} = -\left(\mathbf{i}\frac{\partial}{\partial x} + \mathbf{j}\frac{\partial}{\partial y} + \mathbf{k}\frac{\partial}{\partial z}\right)V = -\boldsymbol{\nabla} V$$

where $\boldsymbol{\nabla} V$ (read "del V") is known as the gradient of the scalar V, and

$$\boldsymbol{\nabla} = \mathbf{i}\frac{\partial}{\partial x} + \mathbf{j}\frac{\partial}{\partial y} + \mathbf{k}\frac{\partial}{\partial z} \tag{10-4}$$

is known as the *gradient operator*. Note that V in this example is a scalar, but $\boldsymbol{\nabla}$ operating on V produces a vector $\boldsymbol{\nabla} V$. Hence, force is the negative gradient of the potential energy.

It is possible to use $\boldsymbol{\nabla}$ in a scalar product with another vector **A**. Such a product $\boldsymbol{\nabla} \cdot \mathbf{A}$ is a scalar called the *divergence* of **A**. If ϕ be any scalar, we find that the divergence of the gradient of ϕ in Cartesian coordinates is

$$\boldsymbol{\nabla} \cdot \boldsymbol{\nabla}\phi = \boldsymbol{\nabla}^2\phi = \frac{\partial^2\phi}{\partial x^2} + \frac{\partial^2\phi}{\partial y^2} + \frac{\partial^2\phi}{\partial z^2} \tag{10-5}$$

The scalar operator $\boldsymbol{\nabla}^2$ (read "del squared") is known as the *Laplacian operator*:

$$\boldsymbol{\nabla}^2 = \frac{\partial^2}{\partial x^2} + \frac{\partial^2}{\partial y^2} + \frac{\partial^2}{\partial z^2} \tag{10-6}$$

10-3 EIGENVALUE EQUATIONS REVISITED

We considered in Chapter 9 a matrix description of a set of linear equations written

$$\mathbf{Ac} = \lambda\mathbf{c}$$

where λ is a set of constants called the eigenvalues of the matrix **A** and **c** is a vector. We recognize the **A** matrix to be an operator, since its effect is to multiply each element in the **c** matrix by the constant λ. We find that eigenvalue equations are not restricted to matrix mathematics and find general use in many areas of applied mathematics. In fact, we can define as an *eigenvalue equation* any operator equation having the general form

$$\hat{\mathbf{A}}\phi = a\phi \tag{10-7}$$

where $\hat{\mathbf{A}}$ is an operator whose operation on a function ϕ, called an *eigenfunction* of the operator, produces a set of constants, the *eigenvalues*, multiplied by the function ϕ.

For example, the differential equation

$$\frac{d}{dx}(e^{mx}) = me^{mx}$$

is an eigenvalue equation.

It is possible for operators to have several eigenfunctions. For example, the functions e^{mx}, e^{-mx}, e^{imx}, e^{-imx}, sin mx, and cos mx are all eigenfunctions of the operator $\hat{\mathbf{D}}_x^2 = \partial^2/\partial x^2$. Some, however, will be unsuitable in a problem because of certain restrictions on the functions, known as *boundary conditions*.

We can solve eigenvalue equations by methods outlined for differential and partial differential equations in Chapter 6. Consider the following examples.

Examples

(a) Show that the functions $\phi = e^{-ar}$ are eigenfunctions of the operator $\hat{\mathbf{D}}^2 = \partial^2/\partial r^2$. What are the eigenvalues?

If $\phi = e^{-ar}$ are eigenfunctions of $\hat{\mathbf{D}}^2$, then they should satisfy the eigenvalue equation

$$\hat{\mathbf{D}}^2\phi = k\phi$$

where the constants k are the eigenvalues. Differentiating ϕ with respect to r, we have

$$\hat{\mathbf{D}}\phi = -ae^{-ar}$$
$$\hat{\mathbf{D}}^2\phi = a^2e^{-ar} = a^2\phi$$

Thus, we see that the functions $\phi = e^{-ar}$ satisfy the eigenvalue equation and that $k = a^2$ are the eigenvalues.

It is important to stress that for a function to be an eigenfunction of an operator, the operator, upon acting on the function, must reproduce the exact same function. Thus, we see that sin mx is not an eigenfunction of the differential operator d/dx, since the operator acting on sin mx produces a set of constants multiplied by cos mx, a different function.

(b) Show that the Schrödinger equation in one dimension

$$\frac{d^2\psi}{dx^2} + \frac{2\mu}{\hbar^2}(E - V)\psi = 0$$

where $\hbar$, μ, and E are constants, is an eigenvalue equation. What are the eigenvalues? (*Hint:* In this equation, both d^2/dx^2 and V are operators.)

Let us begin by putting the Schrödinger equation in eigenvalue equation form $\hat{\mathbf{A}}\phi = a\phi$.

$$\frac{d^2\psi}{dx^2} + \frac{2\mu E\psi}{\hbar^2} - \frac{2\mu V\psi}{\hbar^2} = 0$$

$$-\frac{\hbar^2}{2\mu}\frac{d^2\psi}{dx^2} - E\psi + V\psi = 0$$

or

$$-\frac{\hbar^2}{2\mu}\frac{d^2\psi}{dx^2} + V\psi = E\psi$$

Thus,

$$\left(-\frac{\hbar^2}{2\mu}\frac{d^2}{dx^2} + V\right)\psi = E\psi$$

The operator in parentheses is known as the *Hamiltonian operator* and normally is designated by the symbol $\hat{\mathbf{H}}$. Hence, we can write the Schrödinger equation in eigenvalue form as

$$\hat{\mathbf{H}}\psi = E\psi$$

where the constants E are the eigenvalues.

The Hamiltonian operator illustrates a general property of operators. If $\hat{\mathbf{D}}$ is the sum of two or more operators, $\hat{\mathbf{D}} = \hat{\mathbf{D}}_1 + \hat{\mathbf{D}}_2 + \hat{\mathbf{D}}_3 + \cdots$, then $\hat{\mathbf{D}}f(x) = (\hat{\mathbf{D}}_1 + \hat{\mathbf{D}}_2 + \hat{\mathbf{D}}_3 + \cdots)f(x) = \hat{\mathbf{D}}_1 f(x) + \hat{\mathbf{D}}_2 f(x) + \hat{\mathbf{D}}_3 f(x) + \cdots$.

10-4 HERMITIAN OPERATORS

The eigenvalue problems found in physical chemistry, particularly those in the area of quantum mechanics, are all described by an eigenvalue equation having the general form

$$L(u) + \lambda w u = 0 \tag{10-8}$$

Generally, the operator L is a second-order differential operator that operates on the function u, the constants λ are the eigenvalues, and w is a weighting factor. Second-order differential operators of the form

$$\hat{\mathbf{L}}u = f\frac{d^2u}{dx^2} + g\frac{du}{dx} + hu$$

where f, g, and h are functions of x, are said to be *self-adjoint* if $g = \frac{df}{dx}$. If the differential operator is not self-adjoint, it can be made self-adjoint by multiplying it by a factor $\int \frac{g - \frac{df}{dx}}{f}\,dx$. For example, the Legendre's equation is already self-adjoint. Hermite's equation can be made self-adjoint by multiplying it through by e^{-x^2}.

Self-adjoint operators, such as those described above, are found to obey an important rule: If u and v are two acceptable functions of q, then an operator $\hat{\mathbf{P}}$ is said to be *Hermitian* if

$$\int_{\text{all space}} u^*\hat{\mathbf{P}}v\,d\tau = \int_{\text{all space}} v\hat{\mathbf{P}}^*u^*\,d\tau \tag{10-9}$$

where the * denotes the complex conjugate.

Example

Show that the momentum operator $\hat{\mathbf{p}}_q = -i\hbar(\partial/\partial q)$ is Hermitian.

$$\int u^*\hat{\mathbf{p}}_q v\,dq = -i\hbar\int u^*\frac{\partial v}{\partial q}\,dq$$

Solution. Integrating the right-hand side by parts, we have

$$-i\hbar\int u^*\frac{\partial v}{\partial q}\,dq = -i\hbar\left\{\left[u^*v\right]_{\text{endpoints}} - \int v\frac{\partial u^*}{\partial q}\,dq\right\}$$

We assume that u and v vanish at the endpoints. Thus, the expression in brackets is zero, which gives

$$\int u^*\hat{\mathbf{p}}_q v\,dq = +i\hbar\left\{\int v\frac{\partial u^*}{\partial q}\,dq\right\} = \int v\hat{\mathbf{p}}_q^* u^*\,dq$$

The operator $\hat{\mathbf{p}}_q$ is Hermitian.

A very important property of Hermitian operators is that the eigenvalues of Hermitian operators are always real. To see this, consider the eigenvalue equation

$$\hat{\mathbf{p}}\psi = p\psi \tag{10-10}$$

where $\hat{\mathbf{p}}$ is a Hermitian operator. It also must be true that

$$\hat{\mathbf{p}}^*\psi^* = p^*\psi^* \tag{10-11}$$

We now multiply Equation (10-10) by ψ^*, Equation (10-11) by ψ, and integrate over all space, which gives

$$\int \psi^*\hat{\mathbf{p}}\psi\,d\tau = p\int \psi^*\psi\,d\tau$$

$$\int \psi\hat{\mathbf{p}}^*\psi^*\,d\tau = p^*\int \psi^*\psi\,d\tau$$

Since

$$\int \psi^*\hat{\mathbf{p}}\psi\,d\tau = \int \psi\hat{\mathbf{p}}^*\psi^*\,d\tau$$

Therefore, it must be true that $p^* = p$; the eigenvalues are real.

10-5 ROTATIONAL OPERATORS

We saw in a previous section that operators can take matrix form. In the eigenvalue problem, the matrix operator affects the magnitude or length of a vector, the eigenvector, without changing its direction. Consider, now, another type of matrix operator that changes the direction of a vector, but not its magnitude. Such an operator is called a *rotational operator*, since its effect is to rotate a vector about the origin of a coordinate system by an angle θ. As we shall see, it does this by, in essence, rotating the coordinate system back through the angle θ.

Consider, as a simple example, a vector **r** extending from the origin of a Cartesian coordinate system to the point (x_1, y_1), as shown in Fig. 10-1(a). Suppose that we wish to rotate this vector through an angle θ to another point (x_2, y_2). We can do this either by rotating the vector itself through the angle θ [Fig. 10-1(b)], or by rotating the coordinate system back through the angle θ [Fig. 10-1(c)].

The coordinates (x_1, y_1) can be related to the coordinates (x_2, y_2) by simple trigonometry. We see from Fig. 10-1(b) that

$$\theta = 180° - (\alpha + \beta) = 180° - \alpha - \beta$$

$$x_1 = r \cos \alpha$$

$$y_1 = r \sin \alpha$$

$$x_2 = -r \cos \beta$$

$$y_2 = r \sin \beta$$

$$\beta = 180° - (\theta + \alpha)$$

We now must make use of some important trigonometric identities:

$$\begin{aligned} \sin (A + B) &= \sin A \cos B + \cos A \sin B \\ \sin (A - B) &= \sin A \cos B - \cos A \sin B \\ \cos (A + B) &= \cos A \cos B - \sin A \sin B \\ \cos (A - B) &= \cos A \cos B + \sin A \sin B \end{aligned} \qquad (10\text{-}12)$$

Hence,

$$\begin{aligned} \sin \beta = \sin [180° - (\theta + \alpha)] &= \sin 180° \cos (\theta + \alpha) - \cos 180° \sin (\theta + \alpha) \\ &= \sin (\theta + \alpha) = \sin \theta \cos \alpha + \cos \theta \sin \alpha \end{aligned}$$

However,

$$y_2 = r \sin \beta = \sin \theta\, (r \cos \alpha) + \cos \theta\, (r \sin \alpha)$$

Thus,

$$y_2 = x_1 \sin \theta + y_1 \cos \theta$$

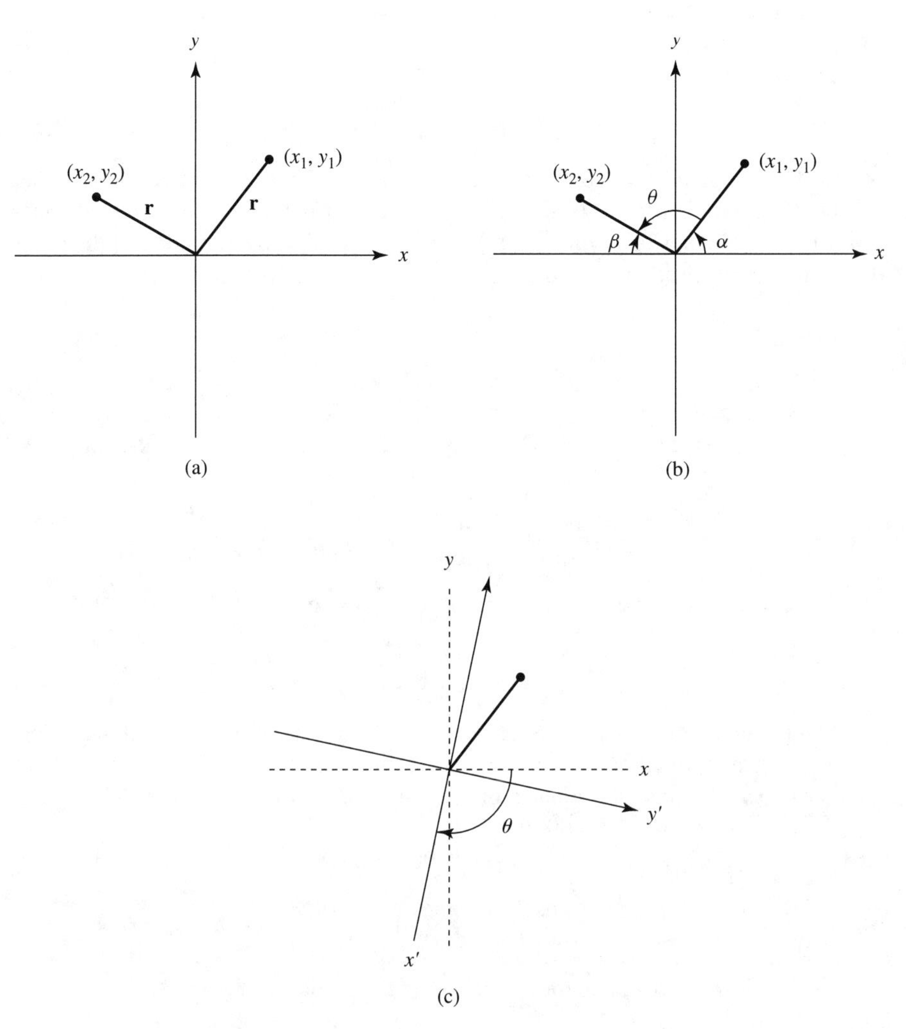

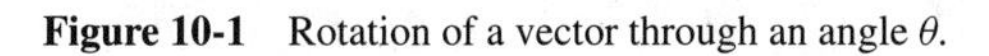

Figure 10-1 Rotation of a vector through an angle θ.

By the same method used above, we can show that

$$x_2 = x_1 \cos\theta - y_1 \sin\theta$$

These equations can be written in matrix form as

$$\begin{pmatrix} x_2 \\ y_2 \end{pmatrix} = \begin{pmatrix} \cos\theta & -\sin\theta \\ \sin\theta & \cos\theta \end{pmatrix} \begin{pmatrix} x_1 \\ y_1 \end{pmatrix} \tag{10-13}$$

where we recognize $\begin{pmatrix} x_1 \\ y_1 \end{pmatrix}$ as the original vector and $\begin{pmatrix} x_2 \\ y_2 \end{pmatrix}$ as the rotated vector.

The matrix

$$\begin{pmatrix} \cos\theta & -\sin\theta \\ \sin\theta & \cos\theta \end{pmatrix}$$

is a rotational operator, since it rotates a vector about the origin in two-dimensional space. This operator also can be thought of as a transformation operator, since it transforms the coordinates of a point (x_1, y_1) into a point (x_2, y_2) by a rotation of the coordinate axes back through an angle θ.

Examples

(a) Rotate the vector $\mathbf{r}(3, 4)$ through an angle of 180° and find the new direction of the vector.

Since sin 180° = 0 and cos 180° = −1, we have for the rotational operator

$$\begin{pmatrix} \cos\theta & -\sin\theta \\ \sin\theta & \cos\theta \end{pmatrix} = \begin{pmatrix} -1 & 0 \\ 0 & -1 \end{pmatrix}$$

Allowing this operator to operate on $\mathbf{r}(3, 4)$ gives

$$\begin{pmatrix} -1 & 0 \\ 0 & -1 \end{pmatrix}\begin{pmatrix} 3 \\ 4 \end{pmatrix} = \begin{pmatrix} -3 \\ -4 \end{pmatrix}$$

which produces a new vector $\mathbf{r}(-3, -4)$. It is easy to verify that the difference in direction between $\mathbf{r}(-3, -4)$ and $\mathbf{r}(3, 4)$ is 180°.

(b) Find the new coordinates of the point (3, 4) after a rotation of the point about the origin by 90°.

Since sin 90° = 1 and cos 90° = 0, we have

$$\begin{pmatrix} \cos\theta & -\sin\theta \\ \sin\theta & \cos\theta \end{pmatrix} = \begin{pmatrix} 0 & -1 \\ 1 & 0 \end{pmatrix}$$

Allowing this operator to operate on the point (3, 4) gives

$$\begin{pmatrix} 0 & -1 \\ 1 & 0 \end{pmatrix}\begin{pmatrix} 3 \\ 4 \end{pmatrix} = \begin{pmatrix} -4 \\ 3 \end{pmatrix}$$

Rotational operators are useful in describing the symmetry of molecules. A designation known as Schoenflies notation uses the symbol $\mathbf{C}_n$ (where $n = 360/\theta$) to denote a particular rotational operator. For example, an operator that rotates a molecule through an angle of 90° is called a $\mathbf{C}_4$ operator. A molecule is said to possess C_4 symmetry if a $\mathbf{C}_4$ operation leaves the molecule unchanged. Similar matrix operators for other types of molecular symmetry also are possible.

Consider the water molecule shown in Fig. 10-2. If we apply a $\mathbf{C}_2$ rotational operator to a water molecule such that it rotates the molecule through an angle of 180°

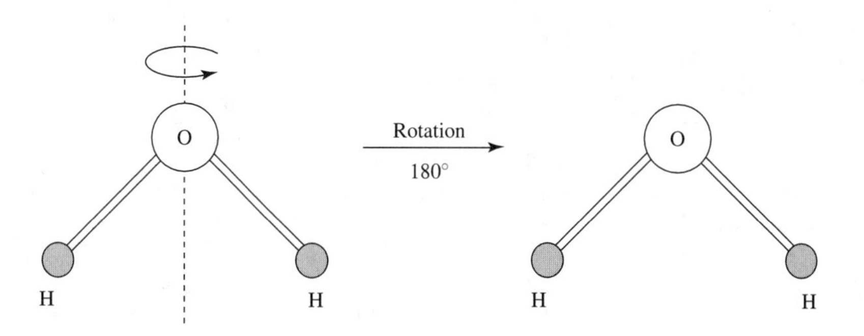

Figure 10-2 Rotation of a water molecule through an angle of 180°.

about the axis shown in the figure (dashed line), the molecule remains unchanged. (The hydrogen atoms are assumed to be indistinguishable). We say, then, that the water molecule possesses a C_2 axis of rotation. By assigning an x- and y-coordinate to each hydrogen atom (the oxygen is placed at the origin), we can easily show this to be the case.

10-6 TRANSFORMATION OF ∇^2 TO PLANE POLAR COORDINATES

Many problems in classical and quantum mechanics require the use of the ∇^2 operator in coordinates other than Cartesian coordinates, the coordinate system in which it has its simplest form. In this section we shall demonstrate the transformation of ∇^2 to plane polar coordinates. We choose plane polar coordinates as an example because, in this transformation, all the essentials of transforming ∇^2 are demonstrated; and yet the transformation is less lengthy than transformation to, say, spherical polar coordinates. For those interested, the essentials of the transformation of ∇^2 to spherical polar coordinates are given in Appendix III.

The Laplacian operator in two dimensions is given by the equation

$$\nabla^2 = \frac{\partial^2}{\partial x^2} + \frac{\partial^2}{\partial y^2}$$

Recall from Chapter 1 that the transformation and reverse transformation equations to plane polar coordinates are

$$x = r\cos\theta \qquad r = (x^2 + y^2)^{1/2}$$

$$y = r\sin\theta \qquad \tan\theta = \left(\frac{y}{x}\right)$$

The transformation of the first derivative operators $\partial/\partial x$ and $\partial/\partial y$ can be found by using the chain rule

$$\frac{\partial}{\partial x} = \frac{\partial r}{\partial x}\frac{\partial}{\partial r} + \frac{\partial \theta}{\partial x}\frac{\partial}{\partial \theta} \quad \text{and} \quad \frac{\partial}{\partial y} = \frac{\partial r}{\partial y}\frac{\partial}{\partial r} + \frac{\partial \theta}{\partial y}\frac{\partial}{\partial \theta} \tag{10-14}$$

We now determine the transformation derivatives from the reverse transformation equations.

$$\frac{\partial r}{\partial x} = \frac{1}{2}(x^2 + y^2)^{-1/2}(2x) = \frac{x}{r} = \frac{r\cos\theta}{r} = \cos\theta$$

Likewise,

$$\frac{\partial r}{\partial y} = \sin\theta$$

Finding $\partial\theta/\partial x$ and $\partial\theta/\partial y$ is a little more involved. A little trick, to keep from having to differentiate the inverse tangent, is to differentiate the tan θ equation rather than solve the equation for θ. Remembering that $d\,(\tan\theta) = \sec^2\theta\, d\theta$, we can write

$$\sec^2\theta\, d\theta = -\frac{y}{x^2}\,dx$$

$$\frac{d\theta}{\cos^2\theta} = -\frac{r\sin\theta}{r^2\cos^2\theta}\,dx$$

$$\frac{\partial\theta}{\partial x} = -\frac{\sin\theta}{r}$$

Likewise, using the same procedure, we find

$$\frac{\partial\theta}{\partial y} = \frac{\cos\theta}{r}$$

The first derivatives are therefore

$$\frac{\partial}{\partial x} = \cos\theta\frac{\partial}{\partial r} - \frac{\sin\theta}{r}\frac{\partial}{\partial\theta} \quad \text{and} \quad \frac{\partial}{\partial y} = \sin\theta\frac{\partial}{\partial r} + \frac{\cos\theta}{r}\frac{\partial}{\partial\theta}$$

To find the Laplacian operator, we must now operate each first derivative operator on itself.

$$\begin{aligned}\frac{\partial^2}{\partial x^2} = \frac{\partial}{\partial x}\frac{\partial}{\partial x} &= \left(\cos\theta\frac{\partial}{\partial r} - \frac{\sin\theta}{r}\frac{\partial}{\partial\theta}\right)\left(\cos\theta\frac{\partial}{\partial r} - \frac{\sin\theta}{r}\frac{\partial}{\partial\theta}\right)\\ &= \cos^2\theta\frac{\partial^2}{\partial r^2} + \cos\theta\frac{\partial}{\partial r}\left(-\frac{\sin\theta}{r}\frac{\partial}{\partial\theta}\right) - \frac{\sin\theta}{r}\frac{\partial}{\partial\theta}\left(\cos\theta\frac{\partial}{\partial r}\right)\\ &\quad - \frac{\sin\theta}{r}\frac{\partial}{\partial\theta}\left(-\frac{\sin\theta}{r}\frac{\partial}{\partial\theta}\right)\end{aligned}$$

To perform the operation $\cos\theta \frac{\partial}{\partial r}\left(-\frac{\sin\theta}{r}\frac{\partial}{\partial\theta}\right)$, for example, the $-\sin\theta$ will pass through the $\partial/\partial r$ operator; however, the $(1/r)(\partial/\partial\theta)$ term must be differentiated as a product. This gives

$$\cos\theta \frac{\partial}{\partial r}\left(-\frac{\sin\theta}{r}\frac{\partial}{\partial\theta}\right) = -\sin\theta\cos\theta\left[\frac{1}{r}\frac{\partial^2}{\partial r\,\partial\theta} - \frac{1}{r^2}\frac{\partial}{\partial\theta}\right]$$
$$= -\frac{\sin\theta\cos\theta}{r}\frac{\partial^2}{\partial r\,\partial\theta} + \frac{\sin\theta\cos\theta}{r^2}\frac{\partial}{\partial\theta}$$

The same thing must be done for the other terms in the $\partial^2/\partial x^2$ equation. This gives

$$\frac{\partial^2}{\partial x^2} = \cos^2\theta\frac{\partial^2}{\partial r^2} - \frac{2\sin\theta\cos\theta}{r}\frac{\partial^2}{\partial r\,\partial\theta} + \frac{2\sin\theta\cos\theta}{r^2}\frac{\partial}{\partial\theta} + \frac{\sin^2\theta}{r}\frac{\partial}{\partial r} + \frac{\sin^2\theta}{r^2}\frac{\partial^2}{\partial\theta^2} \tag{10-15}$$

Using the same procedure, for $\partial^2/\partial y^2$ we have

$$\frac{\partial^2}{\partial y^2} = \sin^2\theta\frac{\partial^2}{\partial r^2} + \frac{2\sin\theta\cos\theta}{r}\frac{\partial^2}{\partial r\,\partial\theta} - \frac{2\sin\theta\cos\theta}{r^2}\frac{\partial}{\partial\theta} + \frac{\cos^2\theta}{r}\frac{\partial}{\partial r} + \frac{\cos^2\theta}{r^2}\frac{\partial^2}{\partial\theta^2} \tag{10-16}$$

Adding Equations (10-15) and (10-16) gives

$$\nabla^2 = \frac{\partial^2}{\partial x^2} + \frac{\partial^2}{\partial y^2} = \frac{1}{r}\frac{\partial}{\partial r} + \frac{\partial^2}{\partial r^2} + \frac{1}{r^2}\frac{\partial^2}{\partial\theta^2}$$

SUGGESTED READING

1. Nagle, R. Kent, and Saff, Edward B., *Fundamentals of Differential Equations and Boundary Value Problems*, 2nd ed., Addison-Wesley Publishing Co., Boston, 1996.

PROBLEMS

1. Perform the following operations:

(a) $\sum_{n=0}^{5} x^n$

(b) $\sum_{n=0}^{5} (-1)^n x^n$

(c) ΔE

(d) $\hat{\mathbf{D}}_x(x^3 y)$, $\hat{\mathbf{D}}_x = \frac{\partial}{\partial x}$

(e) $\hat{\mathbf{D}}_x^2(x^2y^3)$

(f) $\hat{\mathbf{D}}_y\hat{\mathbf{D}}_x(x^4y^3)$

(g) $\hat{\mathbf{D}}_z\hat{\mathbf{D}}_y\hat{\mathbf{D}}_x(x^2y^2z^2)$

(h) $\hat{\mathbf{D}}_x\sum_{n=0}^{5}x^n$

(i) $\prod_{n=0}^{4}x_n!$

(j) $\begin{pmatrix}-1 & 0\\ 0 & -1\end{pmatrix}\begin{pmatrix}a\\ b\end{pmatrix}$

2. Determine whether the following pairs of operators commute:

(a) $\left[\hat{\mathbf{D}}_y, \hat{\mathbf{D}}_z\right]$

(b) $\left[\hat{\mathbf{D}}_x, \sum\right]$

(c) $\left[\hat{\mathbf{D}}_x, \Delta\right]$

(d) $\left[\sum, \sqrt{}\right]$

3. Show that $\nabla\cdot\nabla\phi = \dfrac{\partial^2\phi}{\partial x^2} + \dfrac{\partial^2\phi}{\partial y^2} + \dfrac{\partial^2\phi}{\partial z^2}$.

4. Show that $\nabla(\psi\phi) = \psi\nabla\phi + \phi\nabla\psi$.

5. An interpretation of the *Heisenberg uncertainty principle* is that the operator for linear momentum in the x-direction does not commute with the operator for position along the x-axis. If

$$\hat{\mathbf{p}}_x = -i\hbar\frac{\partial}{\partial x} \quad \text{and} \quad \hat{\mathbf{x}} = x$$

(where $\hbar = h/2\pi$ is a constant and $i = \sqrt{-1}$) represent operators for linear momentum and position along the x-axis, evaluate the commutator

$$[\hat{\mathbf{p}}_x\hat{\mathbf{x}} - \hat{\mathbf{x}}\hat{\mathbf{p}}_x]$$

and show that it does not equal zero. (*Hint:* Apply the operators $\hat{\mathbf{x}}$ and $\hat{\mathbf{p}}_x$ to an arbitrary function $\phi(x)$, keeping in mind that $x\phi(x)$ must be differentiated as a product.)

6. Show that $y = \sin ax$ is not an eigenfunction of the operator d/dx, but is an eigenfunction of d^2/dx^2.

7. Show that the functions $\Phi = Ae^{im\phi}$, where A, i, and m are constants, are eigenfunctions of the operator

$$\hat{\mathbf{M}}_z = -i\hbar\frac{\partial}{\partial\phi}$$

What are the eigenvalues? Take $\hbar = h/2\pi$ to be constant and $i = \sqrt{-1}$.

8. Show that the function

$$\psi = \sqrt{\frac{2}{a}}\sin\frac{n\pi x}{a}$$

where n and a are constants, is an eigenfunction of the Hamiltonian operator in one dimension

$$\hat{\mathbf{H}} = -\frac{\hbar^2}{2m}\frac{d^2}{dx^2}$$

What are the eigenvalues? Take $\hbar = h/2\pi$ and m to be constants.

9. Show that the function $\phi = xe^{ax}$ is an eigenfunction of the operator

$$\hat{\mathbf{O}} = \frac{d^2}{dx^2} - \frac{2a}{x}$$

where a is a constant. What are the eigenvalues?

10. Using the two-dimensional rotational operator

$$\begin{pmatrix} \cos\theta & -\sin\theta \\ \sin\theta & \cos\theta \end{pmatrix}$$

find the new coordinates of the point after rotation through the angle

(a) (2, 2) through 30°.
(b) (4, 1) through 45°.
(c) (−4, −3) through 180°.
(d) (3, 2) through 60°.
(e) (1, −3) through 240°.

11. The BF_3 molecule is a planar molecule, the fluorine atoms lying at the corners of an equilateral triangle. By assigning x- and y-coordinates to each fluorine atom (the boron atom being placed at the origin), show that a two-dimensional $\mathbf{C}_3$ operation perpendicular to the x-y plane and through the boron atom will transform the molecule into itself. (*Hint:* Place one B-F bond along the y-axis.)

12. The differential operator for angular momentum is given by the expression

$$\hat{\mathbf{M}} = -i\hbar(\mathbf{r} \times \nabla)$$

where $\hbar$ is a constant, $\mathbf{r} = \mathbf{i}x + \mathbf{j}y + \mathbf{k}z$ and

$$\nabla = \mathbf{i}\frac{\partial}{\partial x} + \mathbf{j}\frac{\partial}{\partial y} + \mathbf{k}\frac{\partial}{\partial z}$$

Assuming $\hat{\mathbf{M}} = \mathbf{i}\hat{M}_x + \mathbf{j}\hat{M}_y + \mathbf{k}\hat{M}_z$, find the components of this operator $\hat{M}_x$, $\hat{M}_y$, and $\hat{M}_z$.

13. Transform the components of angular momentum $\hat{M}_x$, $\hat{M}_y$, and $\hat{M}_z$, found in Problem 12, to spherical polar coordinates.

14. Derive an expression for the total squared angular momentum operator

$$\hat{\mathbf{M}}^2 = \hat{M}_x^2 + \hat{M}_y^2 + \hat{M}_z^2$$

using the expressions found in Problem 13. Remember that the operator $\hat{M}_x^2$ is $\hat{M}_x$ operating on $\hat{M}_x$, and is not found by merely squaring $\hat{M}_x$ (see Section 10-1).

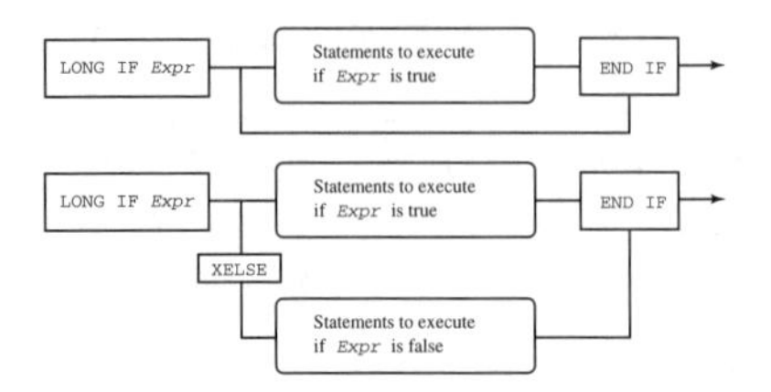

11

Numerical Methods and the Use of the Computer

11-1 INTRODUCTION

During the twenty years since the first edition of this text was written, the computer has become an integral part of the physical sciences and, indeed, a major part of our life in general. Yet, despite its general importance and relevance, including a chapter on computer methods in a text such as this is a risk, because the mathematics found in this text is timeless, while the computer methods are not. There is always the chance of dating a text when it includes subject matter that is so fluid. We find today, however, that most students do not use the computer for the purpose for which it originally was designed—that is, to compute. A primary reason for this is that when personal computers first became available each was hardwired with a programming language, and a major part of learning to use the computer was to learn how to write programs for it. Today most computers, which are much more powerful than those early models, are not supplied with a programming language, and if one wishes to learn how to program the computer, one must purchase a programming language as a separate piece of software.

In this chapter we shall concentrate on how to write programs allowing a computer to do calculations that are particularly useful in physical chemistry. Many students who own computers use them for word processing, spreadsheets, and graphics design. There are many excellent texts available that cover these subjects, so they will not be covered in this chapter. Students are encouraged to continue to use their computers for these purposes, and much spreadsheet software will do some of the types of computing that we will discuss in this chapter. But there is a whole area, new to many students, that should be explored and that is the area of computer programming.

There are a number of programming languages available for programming a computer, and the language chosen depends to a great extent on what the program is intended to do. When the first edition of this text was published, the applicable language for scientific programming was FORTRAN (FORmula TRANslation). With the popular use of the personal computer, the FORTRAN language was replaced by BASIC (Beginners All-purpose Symbolic Instruction Code), and while a number of languages have been developed in recent years to supplant BASIC, it is still one of the best for doing numerical calculations: It is simple and to the point, and it will be the language used in this chapter.

Before we get started discussing programming methods, it is necessary to point out that a computer should not be used as a calculator. Programming a computer to do a single calculation $P = nRT/V$ does not make much sense, since it probably would take less time to punch the data directly into a calculator. Programming a computer to calculate V in van der Waals' equation

$$\left(P + \frac{n^2a}{V^2}\right)(V - nb) = nRT$$

by successive approximations, or to do a Fourier analysis of a wave, does make sense, because these are time-consuming procedures if done on a calculator; the computer can do them in the blink of an eye. So we see that computer program calculations are most useful when they involve reiterative calculations, or when they involve a calculation involving a large number of steps.

As mentioned above, the programming language we are going to use in this chapter is BASIC—in particular, FUTUREBASIC II for Macintosh computers.[1] There are many dialects of BASIC on the market; the particular language will depend on the make and model of the computer being used. In this chapter, every attempt will be made to be as general as possible. Even if you prefer a different language (C and C^{++} are becoming more and more popular), you should be able to make the transformation from BASIC to your language of preference. Also, it is not possible in a text such as this to cover all aspects of programming. You should refer to the particular manual that comes with the programming software. The major purpose of this chapter is to point you in the right direction, and to convince you that writing computer programs is enjoyable and not that difficult.

[1]FUTUREBASIC II is a registered trademark of Staz Software, Inc., 3 Leisure Time Drive, Diamondhead, MS 39525-3215.

11-2 PROGRAMMING A COMPUTER

Variables

A *variable* is an area of memory set aside to hold various kinds of information. (Different kinds of information are called *types*.) Variable names can be letters, words, or numbers, but the names must conform to the following rules. In FUTUREBASIC II the name can be up to 240 characters long, but only the first 15 characters are significant. The name must begin with an alphabetic character. Certain characters are illegal in a name, including: `" , . ˆ ^ + * @ - > = < ] [ ( ) { } ? '`. Also, certain words that are BASIC commands are illegal as variable names. Refer to your manual to get a complete list: some of the common ones are obvious: `SIN`, `COS`, `TAN`, `SQR`, `WHILE`, `IF`, `THEN`, `FOR`, `NEXT`, and so on.

Most BASIC languages use the following symbols to indicate the type of variable:

String Variable. String variables are used to store alphanumeric data. The symbol for a string variable is `$`. A string variable can have a range from 0 to 255 characters (don't confuse the size of the variable with the size of the name). Some examples of string variable names are: `name$`, `x$`, and `element$`. String variables usually describe those variables that are names or words, such as the elements on the periodic chart, the name of a compound, or words. Strings of numbers also can be defined as string variables—`NUM$ = "100"`, or `EXPR$ = "50"`— but numbers defined as string variables cannot be manipulated mathematically. For example, if we were to write in a program, `PRINT NUM$ + EXPR$`, the computer would print `10050`.

Integer Variable. Integer variables are used to describe variables that are integers. Two types of integer variables are possible. The first is simply called "integer," with symbol `%`. An integer `%` has a numerical range of ±32,767. The second type is called "long integer," with symbol `&`. A long integer has a range of ±2,147,483,647. Examples of integer names are: `myList%`, `I%`, `length&`, and so on.

BCD (Binary Coded Decimal) Floating Point Variable. There are at least two floating point precisions that can be configured, when writing programs, to return up to 240 significant digits. The first type is called a single-precision variable, with symbol `!`. Single-precision variables have a range of $10^{\pm 63}$ and, generally, are configured to return a number with 6 significant digits. The second type is called a double-precision variable, with symbol `#`. Double-precision variables have a range of $10^{\pm 16,383}$ and are usually configured to return a number with 12 significant digits. For most calculations, single-precision variables are used, since they use much less memory. For those of you familiar with "bits" and "bytes," the BCD variable memory requirement is (digits + 1)/2 = bytes required per floating point variable. Thus, the greater the number of significant figures asked of the variable, the more memory it uses. Examples of names of single- and double-precision variables are: `x! = 1.111`, `dspacing! = 2.1567`, `pi# = 3.1415926`, and `R! = 0.08206`.

Mathematics

Math Operations. Listed below are some of the common math operations used in BASIC. Some will be familiar to you (being similar to those operations we use in algebra) and some will not.

Operator	*Definition*
+	addition
–	subtraction
*	multiplication
/	division
^	exponentiation
>	greater than
<	less than
>=	greater than or equal to
<=	less than or equal to
<>	not equal to

Examples of the math formats are:

Algebra: $A + 2(C) - 4$ BASIC: `A!+2!*C!-4!`

Note that multiplication and division take precedence over addition and subtraction; the multiplication or division will be performed first.

Algebra: $A\left(\frac{x}{y}\right)$ BASIC: `A!*(x!/y!)`

Algebra: $(C - m) + b^2$ BASIC: `(C! - m#)+ b!^2`

Algebra: $x(-y)$ BASIC: `x!*-y!`

When a number is entered without designating its variable type, most BASIC dialects recognize it as an integer. Thus, an expression `1/4` will produce a value of zero, but `1/4!` will produce a value of 0.25.

Math Functions. Listed below are some of the math functions common to most BASIC dialects:

Function	*Returns*
`INT(expr)`	expression as integer (drops all digits past decimal point without rounding up)

`ABS(expr)`	absolute value of expression
`SIN(expr)`	sine of angle expressed in radians
`COS(expr)`	cosine of angle expressed in radians
`TAN(expr)`	tangent of angle expressed in radians
`ATN(expr)`	arctangent of angle expressed in radians
`SQR(expr)`	square root of absolute value of argument
`LOG(expr)`	natural logarithm of expression
`EXP(expr)`	exponential e^{expr}

Additional special math functions can be found in the various programs described in this chapter.

Scientific Notation. Scientific notation is expressed in BASIC by using `E` for the power of 10. Thus, Avogadro's number would be `6.022E+23`. The Boltzmann constant would be `1.38E-23`. Constants and variables will be expressed in scientific notation when the value is less than 0.01 or greater than 8 digits to the left of the decimal point.

Examples of some of the mathematical functions described here will be given in the special programs listed below (Section 11-3).

Program Statements

Program statements are typed from the keyboard in a manner similar to using a typewriter or word processor. Newer versions of BASIC do not recommend numbering statements, although most will still allow it. More than one program statement can be placed on a line by using a colon; for example,

```
x!=3.0446: Y%=5: Pi#=3.1415926: A!=x!*Y%/Pi#
```

Comment statements can be added in most versions of BASIC and will not be considered by the compiler as part of the program. Most dialects require some type of denotation at the beginning of the statement designating it as a comment statement and not part of the program. Check your programming manual for the proper symbol. In FUTUREBASIC II the symbol is `'`. Examples are:

```
'Program to calculate d-spacings in a powder diffraction pat-
  tern
V!=3.1416!*radius!^2: 'This gives the volume.
  'Main Program
```

Input-Output Statements

Input-output statements are methods of getting your data into the computer and retrieving data from the computer. The common method of introducing data from the keyboard is to use the INPUT statement. When the program reaches an INPUT statement, a question mark prompt will be displayed on the screen. Several pieces of data can be introduced with a single INPUT statement by using commas. Prompt strings also can be specified. For example,

```
INPUT x!
INPUT h,k,l
```

or, with a prompt string (note the syntax),

```
INPUT"Input the Miller indices";h,k,l
INPUT"Do you wish to continue - yes or no?";A$
```

Other methods of introducing data by using loops or from a file will be discussed later in the chapter.

The PRINT statement is the most common method of displaying data on the screen. The PRINT statement can be used with or without a print string. For example,

```
PRINT x!
PRINT h,k,l
PRINT h;k;l
```

or, with a print string (note the syntax),

```
PRINT"The d-spacing is ";dspacing!
PRINT"slope = ";m!, "y -intercept = ";b!
```

When a semicolon is used in the print statement, the cursor on the screen will not move to a new line and will not advance. Subsequent printing continues immediately after the last item printed. When a comma is used, the cursor will not move to a new line but will advance to the next tab stop which can be set with DEF TAB. The default is usually about 5 spaces. A PRINT statement alone causes the cursor to advance to a new line. For example,

```
PRINT name$:PRINT:PRINT address$
```

Data also can be output to a printer. The method by which this is accomplished depends on the dialect of BASIC being used, the computer, and the printer. Older versions of BASIC required actually formatting the printer so that the data would end up at the right position on the page. This is still possible, but is usually not required. Most of the later versions of BASIC recognize the LPRINT statement as the statement to output to the printer. Check your manual—and remember to turn on the printer. One example statement is

```
INPUT"radius of circle";RA!
Area! = 3.1416*RA!*RA!
LPRINT"Area of circle = ";Area!
CLOSE LPRINT
```

Loops

There are many types of loops in computer programming that do a number of things, from simple counting to complex logical routines. The ability of the computer to do calculations that loop back on themselves, to do them quickly, and then to make decisions about these calculations is where the computer shows its greatest power over being a mere calculator. Before discussing various types of loops, let us first consider a major change in the newer forms of BASIC. In the older forms of BASIC, each statement in the program was numbered and GOTO statements could be used to jump around the program at will. This method even allowed one to jump out of a loop before the loop was completed. While the GOTO statements made it easier to design programs, since very little logic in the layout of the program steps had to be followed, it drastically slowed down the execution of the program. Newer BASIC compilers resist the use of GOTO statements in the construction of programs, and some programmers will smugly suggest that the use of GOTO statements in a program is a sign of a weak mind. Thus, in newer forms of BASIC the program steps are not numbered, and the logic of the program (the manner in which the program executes) must be built into the layout of the program steps. This requires a little more time in the program design, but, as a result, the program runs much more quickly and takes advantage of the high speed of the newer computers.

To illustrate how the speed of the computer and the efficiency of the program affect the use of a computer in doing calculations, consider the following personal example. This author has written programs to solve X-ray–diffraction crystal-structure problems using older forms of BASIC on what today is considered "slow," 1-megahertz computers that took as long as a few seconds to do one calculation. At times the program required as many as 40,000 of these calculations; so we would punch in the data and then leave the computer for a few days until it completed all its calculations. We considered ourselves fortunate, since doing the same calculations with a calculator would have taken months. With newer, high-speed computers, and more efficient programs, these same calculations are done in a fraction of second, and the total calculation can be completed in less than an hour. The speed of the computer and the efficiency of the program design *is* extremely important.

FOR-NEXT Loops. One of the simplest loops is a FOR-NEXT loop, essentially a counting loop. The program cycles around the loop a prescribed number of times. While in the loop, however, one can do a large number of things. An example of a simple counting loop is

```
FOR I = 1 TO 20
  PRINT I
NEXT I
```

Executing this little loop will cause the computer to count from 1 to 20.

Other examples are:

(a)
```
J = 0
FOR I=1 TO 20
  J=J+1
  PRINT J
NEXT I
```

This routine illustrates an important difference between algebra and computer logic. The statement `J = J + 1` is algebraically impossible; the computer, however, does not see it as an algebraic expression. Initially, before the loop starts, `J` is defined as zero. A zero is stored in the memory allocated for `J`. When the statement `J = J + 1` comes up, the computer takes the old `J (J = 0)` and adds 1 to it, producing a new `J`, and this new value now is stored in the memory for `J`. Each time the program cycles through the loop, the computer takes the J value stored in the memory and adds 1 to it, producing a new `J`. The `J` counts from 1 to 20.

(b)
```
J!=0
FOR I = 0 TO 10 STEP 2
  J!=J!+1
  J!=SQR(J!)
  PRINT J!
NEXT I
```

Notice here that the loop counts from 0 to 10 in steps of 2. The value of `J`, however, changes by 1. It is also possible to step backward, say, from 10 to 0, using `STEP -`. Also, `J` is defined as a single-precision variable rather than as an integer, since it is the square root of `J` that is being calculated.

It is possible to put loops inside loops. This process is called *nesting*. For example, suppose we wished to print out all the possible permutations of Miller indices `h`, `k`, and `l` from 0 to 6. We can do this with nested `FOR-NEXT` loops. The steps are:

```
FOR h = 0 TO 6
  FOR k = 0 TO 6
    FOR l = 0 TO 6
      PRINT h;k;l
    NEXT l
  NEXT k
NEXT h
```

WHILE-WEND Loops. A `WHILE-WEND` loop is a loop that causes the computer to make a decision. The syntax of a `WHILE-WEND` loop is

```
WHILE Expr
  (Statements here will be executed until Expr is false)
WEND
```

An example of a `WHILE-WEND` loop is:

```
In$="no"
WHILE In$="no"
  INPUT"x? ";x!
  PRINT"Is the value of x correct - yes or no?"
  INPUT In$
WEND
```

The `In$` is defined as `"no"`, and the `WHILE-WEND` loop is executed as long as `In$ = "no"`. After entering a value for `x!`, you are given an opportunity to change `In$`. If you type in `"no"`, you are given the opportunity to change the value of `x!`. If you type in `"yes"`, you exit the loop and the program continues.

Other examples of `WHILE-WEND` loops will appear later in the chapter as parts of programs or routines.

DO-UNTIL Loops. A `DO-UNTIL` loop is very similar to a `WHILE-WEND` loop. The syntax is

```
DO
  (Statements here will be executed until Expr is false)
UNTIL Expr
```

The major difference between a `DO-UNTIL` loop and a `WHILE-WEND` loop is that in the `DO-UNTIL` loop the statements between the `DO` and the `UNTIL` are executed at least once, whereas in the `WHILE-WEND` loop the loop will exit immediately at `WEND` if *`Expr`* is false. An example of a `DO-UNTIL` loop is

```
I = 0
DO
  I = I+1
  PRINT I
UNTIL I = 10
```

The program output will count to 10.

IF-THEN Logic

IF-THEN-ELSE. An `IF-THEN-ELSE` statement is one that causes the computer to decide whether an expression is true or false, but unlike the looping statements above, the decision is made only once. The syntax is

```
IF Expr is true THEN do this [ELSE do this]
```

The ELSE part of the statement is optional. If the ELSE part is not used, the statement simply is called an IF-THEN statement. Some examples of IF-THEN-ELSE statements follow:

(a)
```
FOR I = 1 TO 10
  IF I = 5 THEN PRINT "yes" ELSE PRINT "no"
NEXT I
```

The program output would be:

```
no, no, no, no, yes, no, no, no, no, no
```

(b)
```
flag = 1
IF x<>0 THEN flag=0
```

Closely related to the IF-THEN-ELSE statement is the LONG IF-XELSE-END IF statement. The LONG IF is known as a branching statement. The syntax is

```
LONG IF Expr
  (statements to be executed)
[XELSE]
  (statements to be executed)
END IF
```

Like the IF-THEN-ELSE statement, the XELSE is optional. A flow chart of the statements is shown in Fig. 11-1. An example of the use of a LONG IF statement is

```
xName$="acid"
INPUT"Input the type of compound - acid or base?";cmpd$
LONG IF cmpd$=xName$
  PRINT"The pH of the solution is less than 7."
XELSE
  PRINT"The pH of the solution is greater than 7."
END IF
```

Special Routines

Truncating Numbers. Generally, just as with a calculator, a computer will present numbers to more significant figures than are justified or wanted. It may be necessary to output data variables to only a specific number of significant digits, and

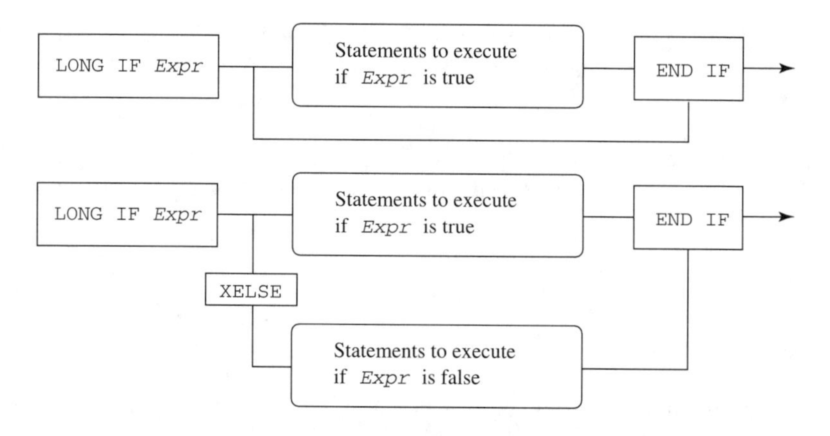

Figure 11-1 Flow chart showing LONG IF-XELSE-END IF logic.

most BASIC dialects have commands that will allow you to truncate a number to a specific number of places past the decimal point. However, there is also a simple little routine that will allow us to do this, illustrated here by a specific example. Suppose we wish to express the variable `A! = 34.778287` to the nearest hundredth (i.e., 2 places past the decimal point). The program steps to do this are:

```
A! = A!*100
A! = A!+0.5
A! = INT(A!)
A! = A!/100
```

The logic behind the steps is as follows. The first step takes `A!` and multiplies it by 100, giving a new `A! = 34.778287 * 100 = 3477.8287`. Because the `INT` statement will not automatically round up a number, we do so artificially by adding 0.5 to produce a new `A!=3477.8287+0.5=3478.3287`. Next, we use the `INT(Expr)` command to change `A!` to an integer `A! = INT(3478.3287) = 3478` (note that the `INT(Expr)` command drops all numbers past the decimal point without rounding up). Finally, we divide `A!` by 100 to produce a new `A!` expressed to the nearest hundredth: `A! = 3478/100 = 34.78`. If we had wanted to express `A!` to the nearest .001, we would first have multiplied by 1000 and then divided by 1000 at the end.

Arrays. An important programming requirement for many scientific-type calculations is the ability to subscript variables, and the use of arrays in computer programming is a way to meet that requirement. The most common variables in scientific calculations are one-dimensional, x_i [syntax `x!(I)`] and two-dimensional, x_{ij} [syntax `x!(I,J)`], but larger multidimensional arrays also are possible. The dimension of the array is always given using a `DIM` statement in one of the first few lines of the program (illustrated below); this allows the computer to set aside memory for the array. The ultimate size of an array will depend on the amount of memory in the computer, since arrays use a considerable amount. Most of the common variable types can be used in arrays. Examples of the syntax are: `A!(10)`, `myArray%(100)`, `Address$(30)`, `x!(4,30)`, `y!(10,6,100)`. Thus, if we wished to define the x and y variables used as, say, 20 points on a graph, these would be one-dimensional arrays `x!(20)`, `y!(20)`. We could input these points with a `FOR-NEXT` loop. The programming steps are:

```
DIM x!(20),y!(20)
FOR I = 1 TO 20
  PRINT I
  INPUT x!(I),y!(I)
  PRINT
NEXT I
```

For a two-dimensional array, we could use nested loops:

```
DIM A!(100,4)
FOR I = 1 TO 50
  FOR J = 1 TO 4
    INPUT A!(I,J)
  NEXT J
NEXT I
```

Sort Routine. It is sometimes useful in programs to be able to arrange a set of numbers in either increasing or decreasing values. Below is a simple little routine that will arrange a set of subscripted variables in decreasing order. With a little modification, the routine can be made to arrange the set in increasing order. Let us assume that we have a set of N variables, W_i, which are randomly distributed and that we wish to arrange these variables in descending order. The routine to do this is

```
FOR J=1 TO (N-1)
  FOR I=1 TO (N-J)
    LONG IF W!(I)<=W!(I+1)
      TE!=W!(I)
      W!(I)=W!(I+1)
      W!(I+1)=TE!
    END IF
  NEXT I
NEXT J
```

We will leave it, as an exercise, for you to explain the logic behind this routine. How would you modify the routine to arrange the variables in ascending order?

Saving Variables to a File. Saving variables to a file is specific to the dialect of BASIC used. Generally, the commands are the same, but you should check the manual for the programming language being used. In FUTUREBASIC II, the commands to save variables to a file are:

```
DIM G!(50)
OPEN "R",1,"myFile"
FOR I=1 TO 45
  WRITE #1,G!(I)
NEXT I
CLOSE #1
```

The first step in the file routine opens a specific file with the name `"myFile"`. The routine will search the folder with the program for this file. If the file does not exist, the routine will create it. The `"R"` designates that this is a *random file* for input and output. Other designations are `"I"` for input only, `"O"` for output only, and `"A"` for append (which positions the pointer at the end of the file). The `"1"` in the `OPEN` statement is the file ID number. This allows more than one file to be opened at the same time. The `WRITE #1` statement indicates that the variables following the `WRITE #1` statement are to be written to file #1. The `CLOSE #1` statement closes file #1. This routine will save the 45 values of `G!` as specific records in a file, named `"myFile"`, in a folder on the disk where the program is found. The `DIM` statement is necessary because `G!` is an array.

To retrieve data from a file, the file must exist in the folder that houses the program. If the file does not exist or is not in the proper folder, the computer will indicate this with an error statement. To retrieve the data stored in the above file, the following statements are used:

```
DIM G!(50)
OPEN "R",1,"myFile"
FOR I=1 TO 45
  READ #1,G!(I)
NEXT I
CLOSE #1
```

It is sometimes not possible to know how many variables are stored in a file and yet necessary to subscript them in an array. To read a file having an unknown number of stored variables, we use the `WHILE (NOT EOF(File ID))-WEND` loop. This will cause the routine to keep reading the file until it runs out of data. An example is

```
DIM symbl$(25), AA!(25)
OPEN "R",1,"scatterFile"
I=0
WHILE (NOT EOF(1))
  I=I+1
  READ #1,symbl$(I);6,AA!(I)
```

```
    DEF TRUNCATE(symbl$(I))
WEND
N=I
CLOSE #1
```

Values for `symbl$` and `AA!` will be read into the computer, each one subscripted with an array number I. Note that the variables `symbl$` and `AA!` had to have been put into the file using a `WRITE #1` statement having exactly the same format as the `READ #1` statement. When the file runs out of variables, the last array number will be designated `N`. Now it will be known how many values of `symbl$` and `AA!` were in the file.

Note also two additional features of the above example. When reading or writing *string* variables from or to a file, it is necessary to indicate the string length (the greatest number of characters that will be in any of the string variables stored). The syntax is `var$; stringLen`. In a `WRITE` statement, the program will reserve this length for every string variable whether it is needed or not. When we read these values back into the program from the file, we do not need all the trailing unused spaces. The `DEF TRUNCATE` command will strip the trailing spaces from the `var$`. For example, in the above routine we reserved six spaces for every `symbl$(I)` entry. Let us say, for example, that one entry is `symbl$(1)="ADDR"` and another is `symbl$(2)="ST"`. Neither uses all six spaces. If we read these back from the file without the `DEF TRUNCATE` statement, `symbl$(1)` will be `ADDR_ _` and `symbl$(2)` will be `ST _ _ _ _`. The `DEF TRUNCATE` statement will drop the trailing spaces, giving `symbl$(1)=ADDR` and `symbl$(2)=ST`.

11-3 PROGRAMMING EXAMPLES

In this section we shall consider a number of programming examples. Some of the programs are quite useful in themselves, but that should not be the major reason for studying this section. Most people learn to program by looking at other programs. You should examine the programs in this section with the idea of picking them apart and using various parts in programs that you are writing. For example, you will see several illustrations of how to introduce your data into a program and a number of examples of loops and how they are used. Pull the programs apart and use the parts. The statements are quite general and common to most BASIC dialects.

Successive Approximations

Let us write a program to calculate V in van der Waals' equation using the method of successive approximations. The van der Waals' equation

$$\left(P + \frac{n^2a}{V^2}\right)(V - nb) = nRT$$

is a cubic in volume. We can approximate a solution to van der Waals' equation by dropping the n^2a/V^2 from the equation and solving the equation for V. This gives

$$V = \frac{nRT}{P} + nb$$

This volume is then used as V in the n^2a/V^2 term in the equation

$$V = \frac{nRT}{\left(P + \frac{n^2a}{V^2}\right)} + nb$$

which again is solved for a new volume V. The process is repeated until there is no appreciable change in the new calculated volume.

The Program. We first input our data as single-precision variables. Recall that the `PRINT` statement after each `INPUT` statement skips a line between each input statement when the program is run. The input statements are:

```
INPUT "number of moles";moles!:PRINT
INPUT "pressure"; press!:PRINT
INPUT "temperature";temp!:PRINT
INPUT"constant a";vanA!:PRINT
INPUT"constant b";vanB!:PRINT
```

Next we perform the calculation. The first calculation finds the initial approximate value of V. Then this value of V is continuously used in a `WHILE-WEND` loop until we are satisfied that the new volume is not significantly different from the previous volume. To keep the `WHILE-WEND` loop going, we set a string variable `A$` to `"YES"`. As long as `A$` is equal to `"YES"` the loop will continue to execute, calculating volumes from previous volumes. Each time, we are given an opportunity to change `A$` to `"NO"`. When we are asked if we wish to continue the loop, we type in either a `"yes"` or a `"no"`. It does not matter whether we type the `"yes"` or `"no"` in uppercase or lowercase letters, because the command `A$ = UCASE$(A$)` always insures that the string variable will be uppercase. When `A$` becomes `"NO"`, we exit the loop.

```
Vol!=(moles!*0.08206*temp!/press!)+moles!*vanB!
A$="YES"
WHILE A$="YES"
  term!=press!+(moles!*moles!*vanA!/(Vol!*Vol!))
  Vol!=(moles!*0.08206*temp!/term!)+moles!*vanB!
  PRINT Vol!
  PRINT "Do you wish to make another determination?"
  INPUT "yes or no?";A$
  A$=UCASE$(A$)
WEND
```

In FUTUREBASIC II the total program will look like this:

```
'Program to Calculate Volume from van der Waals' Equation
WINDOW 1, "van der Waals Volume",(0,0)- (400,300)
INPUT "number of moles";moles!:PRINT
INPUT "pressure";press!:PRINT
INPUT "temperature";temp!:PRINT
INPUT"constant a";vanA!:PRINT
INPUT"constant b";vanB!:PRINT
Vol!=(moles!*0.08206*temp!/press!)+moles!*vanB!
A$ = "YES"
WHILE A$="YES"
  term!=press!+(moles!*moles!*vanA!/(Vol!*Vol!))
  Vol!=(moles!*0.08206*temp!/term!)+moles!*vanB!
  PRINTVol!
  PRINT"Do you wish to make another determination?"
  INPUT"yes or no?";A$
  A$=UCASE$(A$)
WEND
PRINT"Final Volume = ";Vol!:PRINT
INPUT"Press any key to quit";dummy$
END
```

FUTUREBASIC II requires that a window be created in order to display the data. Not all BASIC dialects have this requirement. All data and input prompts are displayed in the window. The size of the window, given by `(0,0)-(400,300)`, is chosen at the programmer's convenience. Also, without the last `INPUT` statement before `END`, the program will not pause after displaying the `"Final Volume"`, but will end immediately and clear the screen.

Linear Regression

One of the best methods for obtaining physical constants from experimental curves is by the *method of least squares*. Although this method can be used for curves of higher order, we shall consider the method only for first-degree or linear equations.

Recall from Chapter 2 that a linear equation is one having the form

$$y = mx + b$$

where m is the slope of the line and b is the y-intercept. Let us assume that for every point (x_i, y_i) in a series of points there is a deviation from linearity due to experimental error of

$$s_i = y_i - mx_i - b$$

The square of this deviation $\sigma_i = s_i^2$ is

$$\sigma_i = y_i^2 - 2mx_iy_i - 2y_ib + m^2x_i^2 + 2mx_ib + b^2$$

and the total squared deviation is the sum of the squared deviations:

$$\sigma = \sum \sigma_i = \sum y_i^2 - \sum 2mx_i y_i - \sum 2y_i b + \sum m^2 x_i^2 + \sum 2mx_i b + \sum b^2 \tag{11-1}$$

The method of least squares requires that the best straight line drawn through the series of points be one in which σ is a minimum with respect to m and b. That is,

$$\frac{\partial \sigma}{\partial m} = \frac{\partial \sigma}{\partial b} = 0$$

Taking the partial derivatives, we have

$$\frac{\partial \sigma}{\partial m} = -\sum 2x_i y_i + \sum 2mx_i^2 + \sum 2x_i b = 0 \tag{11-2}$$

$$\frac{\partial \sigma}{\partial b} = -\sum 2y_i + \sum 2mx_i + \sum 2b = 0 \tag{11-3}$$

Equations (11-2) and (11-3) are two equations that can be solved simultaneously for m and b using Cramer's rule. To set up Equations (11-2) and (11-3) for Cramer's rule, write the equations as

$$m \sum 2x_i^2 + b \sum 2x_i = \sum 2x_i y_i$$
$$m \sum 2x_i + b \sum 2 = \sum 2y_i$$

(Note that the $\sum 2$ means we add 2 for each data point.) Using Cramer's rule, we see that

$$D = \begin{vmatrix} \sum 2x_i^2 & \sum 2x_i \\ \sum 2x_i & \sum 2 \end{vmatrix}, \quad D_1 = \begin{vmatrix} \sum 2x_i y_i & \sum 2x_i \\ \sum 2y_i & \sum 2 \end{vmatrix}, \quad \text{and}$$

$$D_2 = \begin{vmatrix} \sum 2x_i^2 & \sum 2x_i y_i \\ \sum 2x_i & \sum 2y_i \end{vmatrix}$$

$$m = \frac{D_1}{D} \quad \text{and} \quad b = \frac{D_2}{D}$$

The Program. In this program we shall demonstrate another way of introducing data into a program. We begin by inputting the x and y values. The x and y values will be introduced into the program as a one-dimensional array. Remember that most BASIC dialects require one to input the size of arrays at the beginning of the program using a `DIM` statement. This allows the computer to reserve memory for each member of the array. Let us assume that we will never need more than 100 data points in this program. If more are needed, the `DIM` statement can always be changed. The x and y values will be input as single-precision variables. Forgetting to dimension an array will normally result in an `"out of memory"` error statement from the computer. The input statements are as follows:

```
DIM x!(100),y!(100)
PRINT"Input number of data points"
INPUT N
PRINT:PRINT"Input x and y values for each point."
PRINT
FOR I=1 TO N
  PRINT"Input x(";I;"),y(";I;")"
  INPUT x!(I),y!(I)
NEXT I
In$ = ""
WHILE In$<>"YES"
  CLS
  FOR I=1 TO N
    PRINT I,x!(I),y!(I)
  NEXT I
  PRINT:PRINT"Are these values correct - yes or no?"
  INPUT In$
  In$=UCASE$(In$)
  LONG IF In$<>"YES"
    INPUT"Input point number ";I
    INPUT"Input new point values ";x!(I),y!(I)
  END IF
WEND
```

There are several useful programming tips here; let us see if we can sort them out. We first input the number of data points on the graph. Next, using a `FOR-NEXT` loop, we input each x and y value. Examine the `PRINT` statement in the first `FOR-NEXT` loop. Everything within the `""` will be printed as is, and the semicolon will not advance the cursor; so on the screen for `I = 1` it will print as `"Input x(1),y(1)"`. The computer program will assign an I value from 1 to N for each x and y value. If we make an error inputting the data, we continue until all the data has been entered. We now use a `WHILE-WEND` loop to allow us to correct any of the x and y values. The `CLS` statement clears the screen. The program then lists all the x and y values and asks us whether we are satisfied with the data. It gives us an opportunity to change the string variable `In$` to `"YES"`. If we type in `"yes"` or `"YES"`, we will exit the `WHILE-WEND` loop and the program will continue. If we type in anything except `"YES"`, the program will ask us for a point number I and the new values for x_i and y_i. The `WHILE-WEND` loop will recycle, listing all the points, and ask us if we now are satisfied with the data. We can correct as many points as we wish.

The next part of the program deals with the least squares calculation. Let us list the statements first and then explain what they do:

```
A!=0: B!=0: C!=0: H!=0: J!=0
FOR I=1 TO N
  A!=A!+2!
  B!=B!+2!*x!(I)
  C!=C!+2!*x!(I)*x!(I)
  H!=H!+2!*y!(I)
  J!=J!+2!*x!(I)*y!(I)
NEXT I
DET!=A!*C!-B!*B!
bDET!=H!*C!-B!*J!
mDET!=A!*J!-B!*H!
b!=ADET!/DET!
m!=BDET!/DET!
PRINT:PRINT"For the line y = mx+b"
PRINT"m = ";m!
PRINT"b = ";b!
```

We first define five single-precision variables as being equal to zero. Then, using a `FOR-NEXT` loop for the N data points, we determine the summations found in the determinants required for a Cramer's rule solution to simultaneous equations for m and b. For example, as `A!` cycles through the loop, it adds 2 each time for N times. `A!` determines $\sum 2$. Likewise, `B!` determines $\sum 2x_i$, and so on. Once the sums are determined, we calculate the determinants D, D_1, and D_2, which allows us to calculate m and b. The total program in FUTUREBASIC II is:

```
'Program to Find Best Straight Line Through a Set of Points
WINDOW 1, "Least Squares Determination",(0,0)-(400,300)
DIM x!(100),y!(100)
PRINT"Input number of data points"
INPUT N
PRINT:PRINT"Input x and y values for each point."
PRINT
FOR I=1 TO N
  PRINT"Input x(";I;"),y(";I;")"
  INPUT x!(I),y!(I)
NEXT I
In$=""
WHILE In$<>"YES"
  CLS
  FOR I=1 TO N
    PRINT I,x!(I),y!(I)
  NEXT I
  PRINT:PRINT"Are these values correct - yes or no?"
```

```
    INPUT In$
    In$=UCASE$(In$)
    LONG IF In$<>"YES"
      INPUT"Input point number ";I
      INPUT"Input new point values ";x!(I),y!(I)
    END IF
  WEND
  A!=0: B!=0: C!=0: H!=0: J!=0
  FOR I=1 TO N
    A!=A!+2!
    B!=B!+2!*x!(I)
    C!=C!+2!*x!(I)*x!(I)
    H!=H!+2!*y!(I)
    J!=J!+2!*x!(I)*y!(I)
  NEXT I
  DET!=A!*C!-B!*B!
  bDET!=H!*C!-B!*J!
  mDET!=A!*J!-B!*H!
  b!=bDET!/DET!
  m!=mDET!/DET!
PRINT:PRINT"For the line y = mx + b"
PRINT"m = ";m!
PRINT"b = ";b!
INPUT"Press any key to end program.";dummy$
END
```

11-4 NUMERICAL INTEGRATION

In Chapter 5 we defined the geometric interpretation of the definite integral as representing the area under the curve between the limits of integration. There are at least two situations where analytical integration is not possible, and yet it would be useful to know the area under a curve. The first is where the equation of the curve is not known. In these cases, some form of graphical integration (described in Chapter 12) can be performed. The second case is where the equation of the curve is known, but the function cannot be integrated analytically, such as the function $e^{\pm ax^2}$. In these cases, some form of numerical approximation to the area, called *quadrature*, can be used. A large number of numerical methods of integration have been developed, and many of them are particularly suited to the computer. In this section we shall consider a few of the more popular methods.

Integration by the Trapezoid Method

One of the simplest and most straightforward methods of integration involves dividing the interval between the limits of integration into a number of equal subdivisions and then extending vertical lines from the abscissa of the coordinate system to the

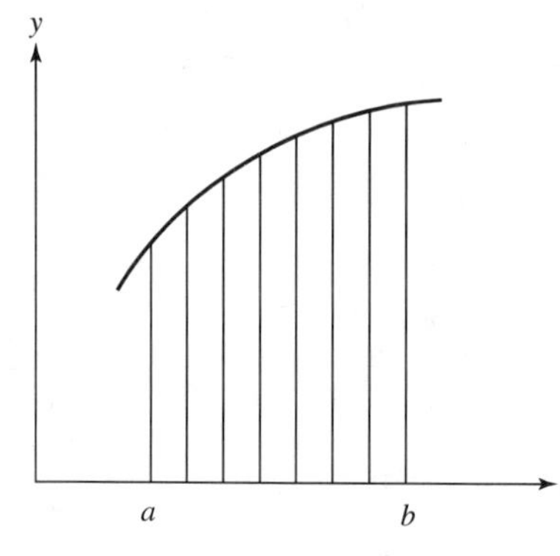

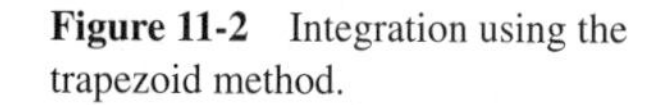
Figure 11-2 Integration using the trapezoid method.

curve. The points where these lines intersect the curve are connected together by straight lines forming a series of trapezoids, as shown in Fig. 11-2. The sum of the areas of the trapezoids closely approximates the area under the curve.

The area of a trapezoid is $A = \frac{1}{2}w(c + d)$, where w is the width of the base and c and d are the lengths of each side. If we assume that $y = f(x)$ represents the equation of the curve, then the area of one of the trapezoids is

$$A = \frac{1}{2}\Delta x[f(x_1) + f(x_2)] \tag{11-4}$$

The sum of the areas of two adjacent trapezoids is

$$\begin{aligned} A &= \Delta x\left[\frac{f(x_1)}{2} + \frac{f(x_2)}{2} + \frac{f(x_2)}{2} + \frac{f(x_3)}{2}\right] \\ &= \Delta x\left[\frac{f(x_1)}{2} + f(x_2) + \frac{f(x_3)}{2}\right] \end{aligned} \tag{11-5}$$

We see, then, that when the curve is divided into n intervals, the total area under the curve between a and b is

$$A = \Delta x\left[\frac{f(a)}{2} + f(x_1) + f(x_2) + \cdots f(x_{n-1}) + \frac{f(b)}{2}\right] \tag{11-6}$$

The Program. Let us now consider a program to determine the definite integral

$$y = \int_0^{\pi} \sin x \, dx = 2.000$$

using the trapezoid method. Note that most BASIC dialects require angles to be expressed in radians. The FUTUREBASIC II program is:

```
'Program for numerical integration using the trapezoid method
WINDOW 1,"Numerical Integration by Trapezoid Method",(0,0)-
   (400,300)
PRINT"Input the limits of integration"
INPUT a!,b!
PRINT
PRINT"Input the number of intervals between limits of integra-
   tion"
INPUT n
DELx!=(b!-a!)/n
FA!=SIN(a!)
FB!=SIN(b!)
x!=a!+DELx!
sum!=0
FOR I=1 TO (n-1)
  y!=SIN(x!)
  x!=x!+DELx!
  sum!=sum!+y!
NEXT I
Area!=DELx!*(sum!+0.5*FA!+0.5*FB!)
PRINT"Area=";Area!
INPUT"Press any key to end";dummy$
END
```

Using 100 intervals, the program produces a value for the area of 1.9998 that is very close to the actual value.

Integration by Simpson's Rule

The trapezoid method approximates the area under the curve by a sum of trapezoids. That is, the points of intersection between the vertical lines and the curve are connected by straight lines. Modifications to the trapezoid method concentrate on how to alter the trapezoids so that they more closely approximate the shape of the curve. The Simpson's rule method utilizes parabolas or second-degree equations to close off the top of each segment. Without proof, the Simpson's rule formula for the summation of the parabolic segments is

$$A = \frac{\Delta x}{3}\Big\{f(a) + f(b) + 4\big[f(x_1) + f(x_3) + \cdots + f(x_{n-1})\big] + 2\big[f(x_2) + f(x_4) + \cdots + f(x_{n-2})\big]\Big\} \quad (11\text{-}7)$$

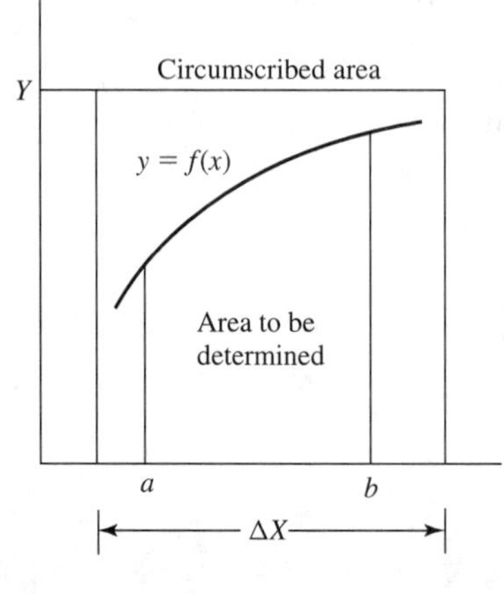

Figure 11-3 Curve illustrating Monte Carlo method.

Note that each $f(x)$ value alternates between being multiplied by 4 and by 2. A program to approximate the integral using Simpson's rule will be very similar to that written for the trapezoid method.

Monte Carlo Method

The Monte Carlo method is a method of numerical integration that is becoming more and more popular as computers become faster. To find the area under a curve using the Monte Carlo method, the area in question is circumscribed by a rectangular area of known value, as shown in Fig. 11-3. Then it is determined whether points (x, y), randomly chosen within this area, fall above or below the curve and within the limits of the integration a and b. The probability that a point falls below the curve is equal to the number of points that actually fall below the curve, divided by the total number of points in the area $Y\ \Delta X$. The area under the curve, then, is

$$\text{Area} = \text{Probability} \times \text{Total rectangular area } Y\ \Delta X$$

This method works well provided a large number (several thousand) of points are tried. Because of the large number of calculations that must be performed, the speed of the computer becomes an important part of the decision whether to use this method of numerical integration.

The Program. The primary part of a program to perform numerical integration using the Monte Carlo method is a random number generator. Most dialects of BASIC have some type of random number generator. Some are true random number generators and some are pseudo-random number generators because the latter generate the same set of "random" numbers each time they are used. In FUTUREBASIC II, the command `RANDOM` at the beginning of the program tells the compiler to use the computer's clock to "seed" the random number generator. In this way, a new set of random numbers is generated each time the random number generator is used.

The command to activate the random number generator is `RND (Expr%)`. The generator will generate an integer from 1 to *Expr%*. For example,

```
RANDOM
FOR J=1 TO 5
  I=RND (100)
  PRINT I
NEXT J
```

The program output would be similar to

```
23,4,44,89,61
```

You should refer to your programming manual to see how your particular random number generator operates.

As in the previous cases, a program to integrate a function numerically using the Monte Carlo method cannot be written entirely in a general form, since the function has to be written into the program each time. Also, the limits of integration and the size of the circumscribed area must be set for each function integrated this way. But a good majority of the program is general enough to be carried over from program to program. In FUTUREBASIC II, a program to integrate a function such as

$$y = \int_0^{\pi/2} \sin x \, dx = 1.000$$

numerically using the Monte Carlo method follows. Note that the limits of integration are from 0 to 1.571 ($\pi/2$ in radians), and that the maximum value of y is sin(1.571) = 1.000. (The fact that the area equals the maximum value of y is accidental.)

```
'Program to Integrate a Function Numerically Using the Monte
  Carlo Method
WINDOW 1,"Numerical Integration Using Monte Carlo Method",
  (0,0)-(400,300)
RANDOM
count=0
FOR K=1 TO 10000
  I=RND (1000)
  xPick!=I/500!: 'This will pick a non-integer x value from 0
   to 2.00
  J=RND(1000)
  yPick!=J/500!: 'This will pick a non-integer y value from 0
   to 2.00
  LONG IF xPick!<=1.571
    y!=SIN(xPick!)
    LONG IF yPick!<=y!
      count=count+1
    END IF
  END IF
```

```
NEXT K
Prob!=count/10000!
Area!=4.00*Prob!:'The circumscribed area is 2 x 2 = 4
PRINT"Probability=";Prob!
PRINT"Area = ";Area!
INPUT"Press any key to end";dummy$
END
```

The total number of points in the circumscribed area is chosen to be 10,000. This is set up in a `FOR-NEXT` loop. Each point, given by `xPick!` and `yPick!`, is found from the random number generator. A circumscribed area of 2.00 × 2.00 will completely surround the sine wave from 0 to $\pi/2$. The first `LONG IF` statement checks to see if the `xPick!` value is within the limits of integration. If it is, the program calculates a true y value. The second `LONG IF` checks to see if the `yPick!` value is above or below this true y value. If it falls below the true value, that point lies in the area under the curve and is counted. After the loop is completed, the count value represents the total number of points under the area. From this, the probability and the area under the curve can be determined. This program, with 10,000 picks, gives an integrated area of 1.02, which is close to the actual area.

11-5 ROOTS TO EQUATIONS

In Chapter 2 we described finding the roots to polynomial equations graphically. Before discussing numerical methods for determining the roots to equations, we should point out that there is a very fast "brute force" computer method of finding these roots; it is not very elegant and is essentially a graphing method. Many computer spreadsheets will perform the standard math and trigonometric calculations found in equations. To find the roots to an equation $y = f(x)$, simply put in a series of x values and let the spreadsheet do its thing. Look for those values of x where the value of y changes sign. Those values of x should be close to the roots of the equation. This procedure can be used with the numerical methods described below to refine the values of x.

There are several numerical methods for findings roots to equation, some better than others. We shall consider only two methods in this section. Those interested in a more complete selection are referred to the readings listed at the end of the chapter.

Method of "Regula Falsi"

The method of "Regula Falsi" is a method of finding a root to an equation (a point where the graph $y = f(x)$ crosses the x-axis) by successive approximations. This method works well in conjunction with, and as a refinement of, the graphical method. Two values of x, say, x_1 and x_2, are selected from the graph on either side of the zero and the corresponding y values are determined. If x_1 is near the desired root, then a better approximation to the root is

$$x = x_1 + \Delta x$$

where

$$\Delta x = \frac{(x_2 - x_1)|y_1|}{|y_1| + |y_2|} \tag{11-8}$$

The process is continued until no further change in the value is obtained within the desired number of significant figures.

Example

Find a root to the equation $y = x^4 + x^3 - 3x^2 - x + 1$ near $x = 0.5$. (A graph of this equation is shown in Fig. 2-9.) Let $x_1 = 0.45$ and $x_2 = 0.50$. Then $y_1 = 0.0746$ and $y_2 = -0.0625$.

$$\Delta x_1 = \frac{(.50 - .45)|.0746|}{|.0746| + |.0625|} = 0.0272; \quad x = 0.45 + 0.0272 = 0.4772$$

A second approximation, with $x_1 = 0.4772$ and $y_1 = 1.65 \times 10^{-4}$, gives essentially no change.

Newton-Raphson Method

The Newton-Raphson method is based on the premise that a line drawn tangent to a curve described by $y = f(x)$ at the point x_1 will intersect the x-axis at a point x_2 having a value closer to the root than x_1. This is illustrated in Fig. 11-4. The slope of a line tangent to a curve at a point x_1 is the derivative $y'(x_1) = dy/dx$. From the diagram we see that

$$y'(x_1) = \tan\theta = \frac{y_1}{(x_2 - x_1)}$$

or

$$x_2 = x_1 + \frac{y(x_1)}{y'(x_1)} \tag{11-9}$$

For the Newton-Raphson method to work, it is necessary that the process converge on the root within a reasonable number of reiterations. This does not always occur; in fact, in some cases we find that the process actually diverges from the root. For this reason, it is advisable to put into any computer program utilizing this method a counter that will automatically terminate the program after some reasonable number of reiterations have been tried.

Example

Find the root to the polynomial $y = x^4 + x^3 - 3x^2 - x + 1$ near $x = -2.5$. First, we must determine $y'(x)$.

$$y'(x) = 4x^3 + 3x^2 - 6x - 1$$

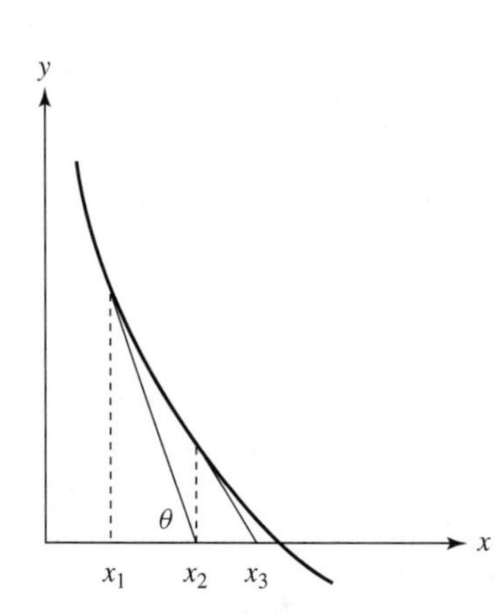

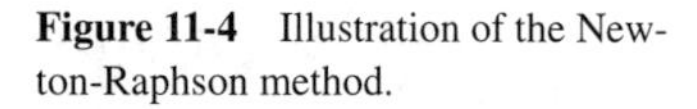
Figure 11-4 Illustration of the Newton-Raphson method.

The first approximation is $x = -2.5$, $y = 8.188$, $y' = -37.25$. Note that, because the initial guess is negative, Equation (11-9) must be modified to $x_2 = x_1 - y(x_1)/y'(x_1)$. Therefore, $x = -2.28$. Using this value for x in the second approximation, we obtain $x = -2.16$. Using this value for x in the third approximation, we obtain $x = -2.10$. Further approximations do not change x appreciably.

SUGGESTED READING

1. BRADLEY, GERALD L., and SMITH, KARL J., *Calculus*, Prentice-Hall, Inc., Upper Saddle River, NJ, 1995.
2. DICKSON, T. R., *The Computer and Chemistry*, W. H. Freeman and Co., San Francisco, CA, 1968.
3. NOGGLE, JOSEPH H., *Physical Chemistry on a Microcomputer*, Little, Brown, and Co., Boston, MA, 1985.
4. VARBERG, DALE, and PURCELL, EDWIN J., *Calculus*, 7th ed., Prentice-Hall, Inc., Upper Saddle River, NJ, 1997.

PROBLEMS

1. Write a program to calculate the average value and the standard error associated with a set of experimental data points (see Chapter 12).

2. Write a program to find the best fit of a second-degree equation, $y = ax^2 + bx + c$, over a set of data points by using the method of least squares. The deviation for each point is $s_i = y_i - ax_i^2 - bx_i - c$, which will lead to solving three simultaneous equations using Cramer's rule.

3. Consider the one-dimensional Fourier series given below:

$$F_0 = +52.0 \qquad F_3 = +25.8$$

$$F_1 = -20.0 \qquad F_4 = -8.9$$

$$F_2 = -14.5 \qquad F_5 = -7.2$$

For a centrosymmetric wave (a wave that is symmetrical about the region of space in which it exists), the Fourier series is

$$f(x) = F_0 + \sum F_n \cos 2\pi nx$$

Write a computer program that will produce values of $f(x)$ from $x = 0$ to $x = 1$ in steps of 0.01. Remember that most BASIC dialects require angles to be in radians.

4. Using the program for numerical integration by the trapezoid method, write a program for numerical integration using Simpson's rule. Try your program on some known functions that can be integrated analytically, such as $y = x^2$.

5. Using the Monte Carlo method, determine the integral

$$A = \int_{-1}^{+1} e^{-x^2}\, dx$$

Compare your answer to that found by using the trapezoid method or Simpson's rule.

6. Integrate the following using the trapezoid method, Simpson's rule, and the Monte Carlo method and compare the results. Express your answers to at least 4 significant figures.

(a) $\int_0^4 \sqrt{x+1}\, dx$ **(c)** $\int_0^{\pi/2} \cos x\, dx$

(b) $\int_1^4 \frac{1}{x}\, dx$ **(d)** $\int_1^2 e^{-x}\, dx$

7. Write a program to find the root of an equation using the method of "Regula Falsi."

8. Write a program to find the root of an equation using the Newton-Raphson method.

9. In the program for calculating V in van der Waals' equation by successive approximations (see the "Successive Approximations" subsection of Section 11-3), the program was terminated by issuing a "yes" or "no" command from the keyboard. Modify the program so that the program itself will terminate automatically when a new value of V is not appreciably different from the previous value of V.

10. The molar heat capacity, C_p, of silver metal is found at temperatures from 20K to 100K to be:

C_p (J · mol^{-1} · K^{-1})	T(K)	C_p (J · mol^{-1} · K^{-1})	T(K)
1.672	20	16.29	70
4.768	30	17.91	80
8.414	40	19.09	90
11.65	50	20.17	100
14.35	60		

Find the change in the enthalpy of silver from 20K to 100K using the trapezoid method, given

$$\Delta H = \int_{20}^{100} C_p \, dT$$

11. The molar heat capacity, C_p, of platinum metal is found at temperatures from 5K to 30K to be:

C_p (J · mol^{-1} · K^{-1})	T(K)
0.1014	5
0.2185	10
0.6438	15
1.444	20
2.673	25
4.136	30

Find the change in the entropy of platinum metal from 5K to 30K using Simpson's rule, given

$$\Delta S = \int_{5}^{30} \frac{C_p}{T} \, dT$$

12. From the data given in the paper by J. O. Hutchens, A. G. Cole, and J. W. Stout, *J. Amer. Chem. Soc.*, **82**, 4813 (1960), determine S^0 for crystalline l-alanine by numerical integration, given

$$S^0 = \int_{0}^{298} \frac{C_p}{T} \, dT$$

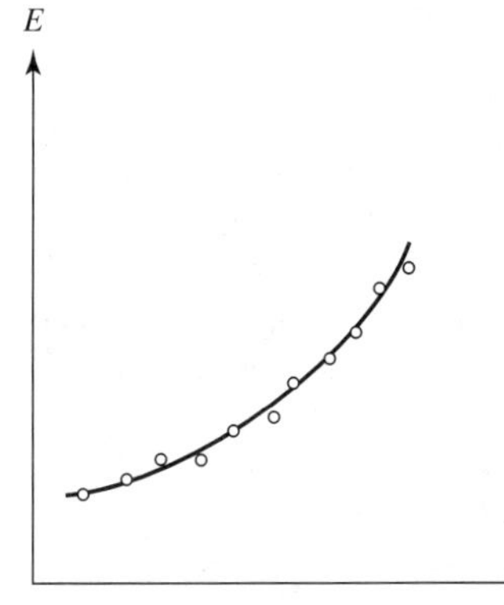

12

Mathematical Methods in the Laboratory

12-1 INTRODUCTION

Most physical chemistry laboratory experiments are concerned with measurements. There is a basic difference, however, between experiments performed at the undergraduate level and those performed in physical chemistry research laboratories. At the undergraduate level, students perform experiments that have a known outcome. Generally, these experiments have been performed many times over a number of years by numerous students. In research laboratories, on the other hand, scientists usually perform experiments on unknowns. There are no laboratory instructors from whom a research scientist can obtain the correct answer to an experimental measurement to see if he or she has performed the measurement correctly. Thus, it is important for students of physical chemistry, who hope someday to become proficient researchers, to learn how to determine the reliability of their experimental data. One common way to help determine the reliability of experimental data is to perform the experiment more than once. It is known that when a measurement is made more than once, the results scatter around some average value. We shall see in the next few sections that this experimental scatter can be used to help determine the probability that the average value is the "true" value. Before going into this, however, let us first review simple probability theory.

12-2 PROBABILITY

The *probability* of any item having a specific characteristic is the number of items having that characteristic divided by the total number of items in the assembly. For example, if we had a barrel containing 50 apples, 40 apples being red and 10 apples being green, the probability of picking a green apple from the barrel (assuming an even distribution) is

$$p = \frac{10}{50} = 0.20 \quad \text{or} \quad 20\%$$

Let us relate this definition of probability to a hypothetical series of measurements. Suppose that we measure on a beam balance the mass of a block of aluminum and that we perform this measurement over and over for a total of 50 measurements. In this example, we are going to assume that the actual mass of the metal does not change (i.e., pieces do not break off, nor does the block get dirty from handling); assume also, to illustrate a point, that the measurements spread over a larger range than we would most probably see if we used a good laboratory balance. The results of the 50 measurements are listed in Table 12-1. These data also are plotted as bar graphs in Fig. 12-1.

Note that the general shape of the graphs does not depend on whether n_i or n_i/N is plotted versus m_i. We find, however, that the shape of the graphs, particularly the height of the bars, does depend on the difference between the measured values Δm. For example, if we had measured the mass to the nearest 0.005 grams rather than to the nearest 0.01 grams, the 50 measurements would have been distributed over 13 data points (from 8.810 to 8.870) rather than over the 7 data points, causing the height of each bar to be reduced.

To make the graphs more comparable, then, let us plot $n_i/(\Delta m \cdot N)$ rather than n_i/N versus m_i. Such a graph is illustrated in Fig. 12-2. Since the shape of this graph no longer depends on Δm, let us allow Δm to approach zero and greatly increase the

TABLE 12-1 MASS OF BAR OF ALUMINUM METAL

Measured Value, m_i (g)	Frequency, n_i	Relative Frequency, n_i/N
8.81	2	0.04
8.82	2	0.04
8.83	11	0.22
8.84	17	0.34
8.85	14	0.28
8.86	3	0.06
8.87	1	0.02

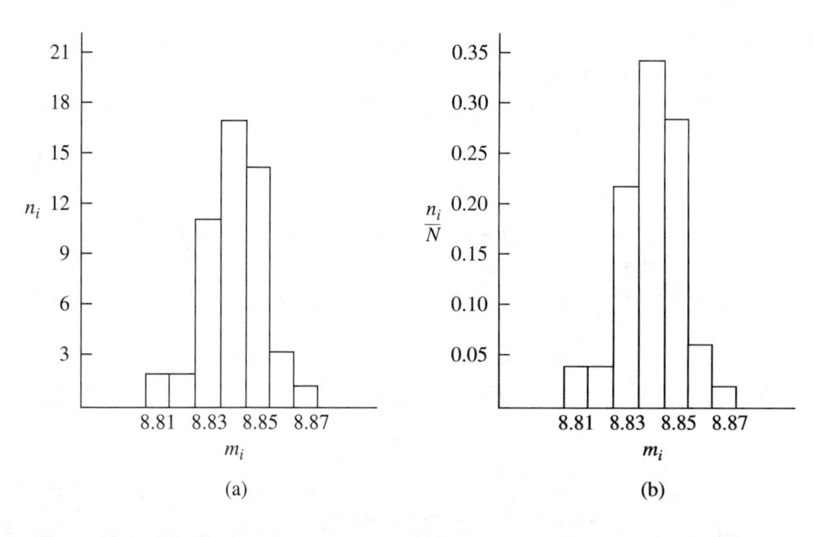

Figure 12-1 Distribution of measurements: (a) frequency versus measured values; (b) relative frequency versus measured values.

number of measurements, N. Hence, we can write

$$\lim_{\Delta m \to 0} \frac{1}{N} \frac{n_i}{\Delta m} = \frac{1}{N} \frac{dn_i}{dm} = P(m) \tag{12-1}$$

where dn_i is the number of measurements lying between m_i and $m_i + dm$. The function $P(m)$ is represented by the dashed line in Fig. 12-2. According to our definition

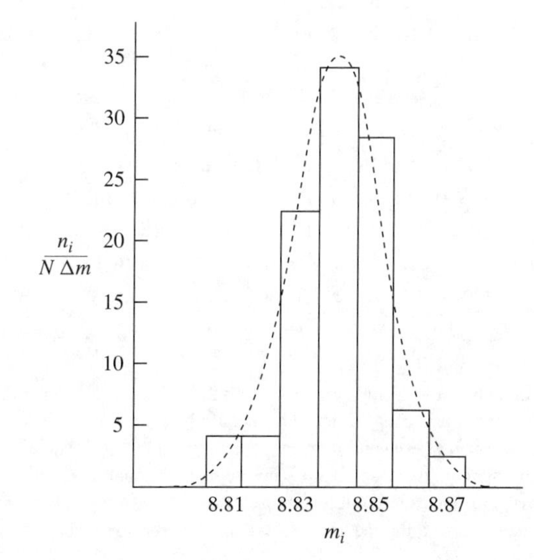

Figure 12-2 Distribution of measured values showing the error probability function $P(m)$.

of probability, the probability that any measurement will have a value lying between m_i and $m_i + dm$ is equal to the number of measurements lying in that range between m_i and $m_i + dm$, which is dn_i, divided by the total number of measurements N.

$$\text{Probability} = \frac{dn_i}{N} = \left(\frac{1}{N}\frac{dn_i}{dm}\right) dm = P(m)dm \tag{12-2}$$

Thus, the function $P(m)$ is called the *error probability function* and represents the distribution of the probability over the measured values. Such a distribution is called a *probability aggregate*. Every measured value is an *element* of that aggregate.

12-3 EXPERIMENTAL ERRORS

Let us now relate what we found in the previous section to experimental errors. As we said earlier, when a measurement is made more than once, the results scatter about an average or mean value. We can define the mean value as

$$\overline{m} = \frac{1}{N}\sum m_i \tag{12-3}$$

This scatter is due to at least two types of error. The first type of error is called *random error* and is due to the fact that various measuring devices have inherent limits in the *precision*[1] to which they can be read. The second type of error is called *systematic error* and is due to such things as uncalibrated instrumentation, human reaction times, and so on.

While it is possible to discuss the random errors associated with measurements, in reality we cannot determine what they actually are. We can, however consider how far each measured value is from the mean value. We define this deviation from the mean, called a *residual*, as

$$r_i = m_i - \overline{m} \tag{12-4}$$

If the errors are only random errors and a large number of measurements are taken, the measurements should fall on a normal distribution curve, such as one described in the previous section. We will show that under these conditions the mean value is the *best value* with a higher precision than any single value. If the errors encountered are mainly systematic errors, no statistical treatment of data can compensate for them. When the errors are entirely random errors, the error probability

[1]The *precision* of a measurement is the relative error between that measurement and other similar measurements performed with the same measuring device. This should be contrasted to the *accuracy* of the measurement, which is the relative error between that measurement and the "true" value of the quantity being measured. It should be pointed out that a set of very precise measurements is not necessarily accurate. For example, if a student were to measure the temperature of a constant-temperature bath several times using a thermometer that had not been calibrated, the results could be very precise but, since the thermometer was not calibrated, very inaccurate.

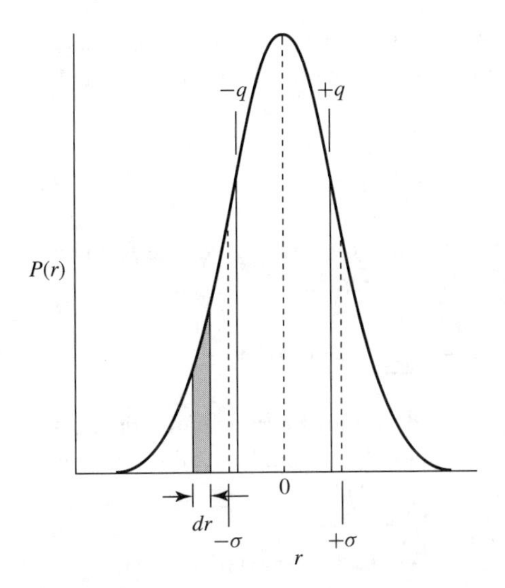

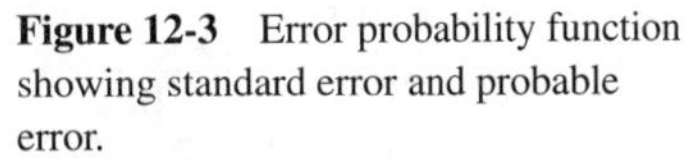
Figure 12-3 Error probability function showing standard error and probable error.

function, plotted as a function of residual, r, rather than measured value, has the form

$$P(r) = \frac{1}{\sigma\sqrt{2\pi}} e^{-r^2/2\sigma^2} \tag{12-5}$$

where σ is a parameter known as the *standard error* or *estimated standard deviation*. In terms of the residuals, the standard error has the form

$$\sigma = \sqrt{\frac{1}{N-1}\sum r_i^2} \tag{12-6}$$

Note that the probability that a single measurement will have a deviation from the mean lying in the range between r and $r + dr$ is $P(r)\,dr$, and this is just the area of the small rectangle of width dr shown in Fig. 12-3. Since the total area under the probability distribution curve must represent the probability of finding the error lying between $-\infty$ and $+\infty$, and this must be 100% or 1, we can write

$$\int_{-\infty}^{+\infty} P(r)dr = 1 \tag{12-7}$$

Equation (12-7) is called the *normalization equation*, since it insures that this integral is equal to unity. If we assume that the error in a single measurement and the residual are essentially the same, then the probability that a measurement will have associated with it an error no larger than $\pm\sigma$ is found to be 68.3% of the total area under the probability curve (illustrated in Fig. 12-3).

Of the many types of parameters used to express uncertainty due to random error, the *probable error* is the most popular. The probable error, q, is defined as having a value such that the probability that an error in a measurement chosen at random will be less than q is equal to the probability that it will be greater than q. That is, there is a 50% probability that the error will lie in the range from $-q$ to $+q$. Stated mathematically,

$$\int_{-q}^{+q} P(r)dr = \frac{1}{2} \tag{12-8}$$

This integral, which can be evaluated by means of a series, gives the following expression for the probable error of a single measurement

$$q = 0.675\sqrt{\frac{1}{N-1}\sum r_i^2} = 0.675\sigma \tag{12-9}$$

Let us evaluate the standard error and probable error for the data given in Table 12-1. The mean value of the 50 measurements is 8.84. Using this value, we find the 50 residuals and from them determine $\Sigma r_i^2 = 0.0072$. This gives

$$\sigma = \sqrt{\frac{1}{N-1}\sum r_i^2} = \sqrt{\frac{1}{49}(0.0072)} = 0.012$$

and

$$q = 0.675\sigma = 0.008$$

Another of the various parameters commonly used by chemists to measure the reliability of a measurement is called the *average error*

$$\bar{r} = \frac{1}{N}\sum |r_i| \tag{12-10}$$

Its advantage is that it is much easier to calculate than is the standard deviation or the probable error.

We said above that when errors are mainly random errors, the mean value is the best value with a higher precision than that of any single measurement. We find that the error associated with the mean value, Q, is inversely proportional to the square root of the number of measurements taken:

$$Q = q/\sqrt{N} \tag{12-11}$$

Thus, the probable error of the mean is

$$Q = 0.675\sqrt{\frac{1}{N(N-1)}\sum r_i^2} \tag{12-12}$$

Applying this equation to the set of data given above, we find

$$Q = \frac{0.008}{\sqrt{50}} = 0.001$$

Hence, we see that while any single measurement in the series is reliable to ± 0.01 g, the mean value has a reliability to ± 0.001 g.

$$\overline{m} = 8.840 \pm 0.001 \text{ g}$$

12-4 PROPAGATION OF ERRORS

Experiments in physical chemistry rarely involve a direct measurement of the physical property in question. For example, in order to determine the density of an object, we normally measure the mass and the volume of the object and relate these to the density. In fact, in many cases we even do not measure the volume of the object directly. Thus, in experiments such as these, the errors are associated with the measured values, the mass and the volume, and not with the density. Yet, it would be useful to know the reliability of the density, given specific errors in the mass and volume.

Suppose that some physical property P is a function of several measurable quantities $x_1, x_2, x_3, \ldots$.

$$P = f(x_1, x_2, x_3, \ldots) \tag{12-13}$$

Further suppose that each of these quantities, measured once or a number of times to determine a mean value, has an error Q associated with it. Let Q_1 be the probable error associated with x_1, Q_2 be the probable error associated with x_2, and so on. We saw in Chapter 4 that any small change in the variables $x_1, x_2, x_3, \ldots$ will cause a change in P given by the equation

$$dP = \left(\frac{\partial P}{\partial x_1}\right) dx_1 + \left(\frac{\partial P}{\partial x_2}\right) dx_2 + \cdots \tag{12-14}$$

Let us assume that the small changes in the variables are the probable errors in these variables. Hence, the probable error in P, $\mathbf{Q}_p$, can be expressed as

$$\mathbf{Q}_p = \left|\frac{\partial P}{\partial x_1}\right| Q_1 + \left|\frac{\partial P}{\partial x_2}\right| Q_2 + \cdots \tag{12-15}$$

where the partial derivatives are evaluated at the mean values, if the x variables are measured more than once.

We find, however, that Equation (12-15) does not take into account that errors in a series of variables tend to cancel each other out. If we square Equation (12-15), we obtain the equation

$$\mathbf{Q}_p^2 = \left(\frac{\partial P}{\partial x_1}\right)^2 Q_1^2 + \left(\frac{\partial P}{\partial x_2}\right)^2 Q_2^2 + \cdots + 2\left(\frac{\partial P}{\partial x_1}\right)\left(\frac{\partial P}{\partial x_2}\right) Q_1 Q_2 + \cdots \tag{12-16}$$

Because probable errors tend to cancel each other out, the cross-terms in Equation (12-16) vanish, giving the equation

$$\mathbf{Q}_p^2 = \left(\frac{\partial P}{\partial x_1}\right)^2 Q_1^2 + \left(\frac{\partial P}{\partial x_2}\right)^2 Q_2^2 + \cdots$$

or

$$\mathbf{Q}_p = \sqrt{\left(\frac{\partial P}{\partial x_1}\right)^2 Q_1^2 + \left(\frac{\partial P}{\partial x_2}\right)^2 Q_2^2 + \cdots} \tag{12-17}$$

where each partial derivative is evaluated at the mean if the x variables are measured more than once. Equation (12-17) allows us to propagate errors through the equations used to calculate the final result.

To further illustrate the use of Equation (12-17), consider the following example. Suppose we measure the mass of a liquid delivered by a 10 ml pipette, using a stoppered weighing bottle and an analytical balance. Let us assume that the probable error in any mass determined on this particular balance is ± 0.0005 g. Assume that the mass of the weighing bottle is 19.9314 ± 0.0005 g and that the mass of the weighing bottle and liquid is 28.1038 ± 0.0005 g. The mass of the liquid therefore is

$$m_l = m_{l+b} - m_b = 28.1038 - 19.9314 = 8.1724 \text{ g}$$

Since the mass of the liquid was not measured directly, we must determine the probable error in the mass of the liquid using Equation (12-17).

$$\mathbf{Q}_{m_l} = \sqrt{\left(\frac{\partial m_l}{\partial m_{l+b}}\right)^2 Q_{m_{l+b}}^2 + \left(\frac{\partial m_l}{\partial m_b}\right)^2 Q_{m_b}^2}$$

Evaluating the partial derivatives

$$\left(\frac{\partial m_l}{\partial m_{l+b}}\right) = 1 \quad \text{and} \quad \left(\frac{\partial m_l}{\partial m_b}\right) = -1$$

Thus, we have

$$\mathbf{Q}_{m_l} = \sqrt{(1)^2(0.0005)^2 + (-1)^2(0.0005)^2} = \pm 0.0007$$

and

$$m_l = 8.1724 \pm 0.0007 \text{ g}$$

Note that the error in the mass of the liquid is larger than the errors in the mass of bottle plus liquid and in the mass of the bottle. This is reasonable, since the mass of the liquid was found by performing two measurements on the balance. On the other hand, the error in the mass of the liquid is less than the sum of the errors in the mass of the bottle plus liquid and mass of bottle. This shows that errors are not purely additive and tend to cancel each other out.

We now determine the density of the liquid by dividing the mass by the volume. Let us assume that the error in the 10 ml pipette is ± 0.02 ml. The density of the liquid is

$$D = \frac{m}{V} = \frac{8.1724}{10.00} = 0.81724 \text{ g/ml}$$

What is the error in the density? To how many significant figures are we justified in expressing the density? Using Equation (12-17), we have

$$\mathbf{Q}_D = \sqrt{\left(\frac{\partial D}{\partial m}\right)^2 Q_m^2 + \left(\frac{\partial D}{\partial V}\right)^2 Q_V^2}$$

Evaluating the partial derivatives,

$$\left(\frac{\partial D}{\partial m}\right) = \frac{1}{V} \quad \text{and} \quad \left(\frac{\partial D}{\partial V}\right) = -\frac{m}{V^2}$$

Thus, we can write

$$\mathbf{Q}_D = \sqrt{\left(\frac{1}{V}\right)^2 (0.0007)^2 + \left(-\frac{m}{V^2}\right)^2 (0.02)^2}$$

$$\mathbf{Q}_D = \sqrt{\left(\frac{1}{10.00}\right)^2 (0.0007)^2 + \left(-\frac{8.1724}{(10.00)^2}\right)^2 (0.02)^2} = \pm 0.002$$

Therefore, the density of the liquid can be expressed as

$$D = 0.817 \pm 0.002 \text{ g/ml}$$

12-5 PREPARATION OF GRAPHS

We saw in Chapter 2 that one of the most useful ways to display the dependence of one function upon another is to use a graph. Many computer spreadsheets have excellent graphing capabilities, and you are encouraged to use them whenever possible. (Computer methods are discussed in Chapter 11.) However, in many situations you still will have to prepare graphs "from scratch," and in doing so, you should use the following approach. First, start with a good grade of graph paper. Engineering graph paper divided 10 × 10 to the centimeter is suitable for most physical chemistry experiments. Other types of graph paper, such as semilog (one axis linear, one axis logarithmic) and log-log (both axes logarithmic) also are available.

Choose a suitable set of coordinate axes and draw these in over lines presently on the graph paper. Be sure to clearly mark and number the main divisions along each axis and to choose a correct scale so that the data does not run off the graph. Label each axis and include the units of the measured values in parentheses after the label—for example, P (atm), V (cc), or T (°C). It is customary also to include a title that

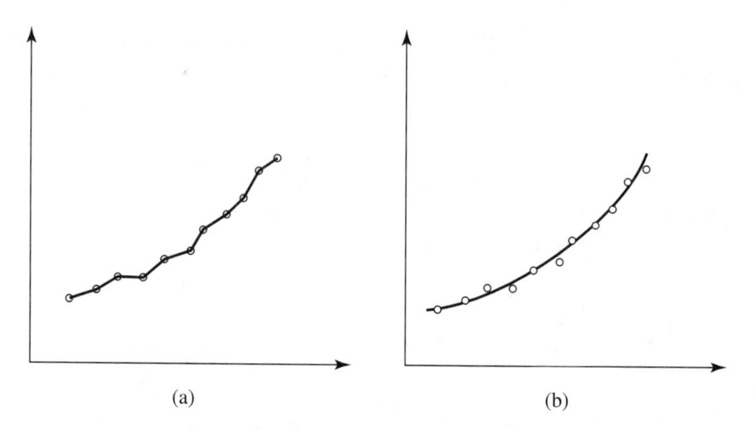

Figure 12-4 Graphical representation of a typical set of data points: (a) incorrect method of connecting points; (b) correct method of drawing smooth curve through a set of points.

describes the data being plotted (for example, *Pressure versus Volume for* CO_2 *Gas*), or for the equation of the curve being plotted (for example, $\ln k = -E_a/RT + \ln A$). The experimental data points should be plotted on the graph paper using a *sharp*, hard-lead pencil or a pin. Circle each point for clarity. It is customary to choose the size of the circle to represent the uncertainty in the experimental point (this, however, is not always convenient).

Once the set of points has been plotted, you then must decide how they should be connected together. It is tempting, for the inexperienced student, to connect the points together by a series of short, straight lines ("follow the dots"), as shown in Fig. 12-4(a). You must keep in mind, however, that experimental data normally are continuous; hence, the data points should be connected by drawing a *smooth* curve through them. This should be done first by sketching lightly through the points with a hard-lead pencil. Do *not* sketch these curves freehand. If the set of points is supposed to be linear, use a straightedge to draw the line; if the set is supposed to fall on a curve, use a French curve, a plastic device containing several irregular curves that can be used as a template. It is not necessary that the curve pass through all the points. It is more important that the curve follow smoothly the trend of the points, as illustrated in Fig. 12-4(b). Do not be afraid to erase if you are not satisfied with the curve. Once the best curve has been drawn through the points, then you may trace over the curve with ink (again using the straightedge or French curve) and erase the penciled lines with a soft eraser. Some laboratory texts suggest not passing the curves through the circles, but only up to their outer edge, since the curve will obscure the experimental points.

12-6 TANGENTS AND AREAS

We find that many experiments in physical chemistry require us to extract information from the slopes of curves. Moreover, it is not always possible to plot data in such

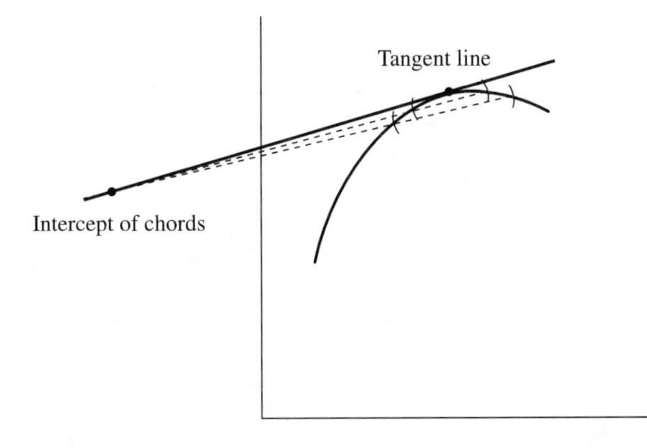

Figure 12-5 Construction of tangent line by method of chords.

a way that straight lines will result (which, of course, would make the determination of the slope very easy). Recall that the slope of a curve at a point is the slope of a line drawn tangent to the curve at that point. If the equation of the curve is known, we can determine the slope from its first derivative. If the equation is not known, then the tangent must be constructed or determined by some numerical method using a computer. (Numerical methods are discussed in Chapter 11.)

One method of constructing tangents to curves is known as the *method of chords*. With a compass placed at the point in question on the curve, strike off two arcs on either side of the point, as shown in Fig. 12-5. Next, draw chords through the intersection of the arcs with the curve and extend the chords until they intersect. A line drawn from the intersection of the two chords to the point on the curve is a good approximation of the tangent to the curve at that point.

It also may be necessary in some experiments to determine graphically the areas under curves. Generally, this is the case in spectroscopy experiments where the intensity of a spectral line is equal to the area under the curve. It is commonly done in gas chromatography also, where the area under the chromatographic peak is directly proportional to the amount of substance causing the peak. There are several acceptable ways to do this. One method, known as the Riemann sum approximation, is to divide the area into a series of rectangles, as shown in Fig.12-6, and measure the area of each rectangle. The rectangles are chosen so that the small triangular areas above the curve approximately equal the small triangular areas below the curve.

Another method, which is quite accurate if done correctly, is to cut out the area and determine its mass on an analytical balance; then this mass is compared to the mass of a known area on the same graph paper. For this method to work, it is important that a good grade of graph paper, having a uniform thickness, be used.

A quick method for determining the area under a curve is to use a planimeter, an instrument designed to give the area when the planimeter is mechanically run about the boundaries of the area. Although this method is fast, it is not very accurate if the area to be determined is small.

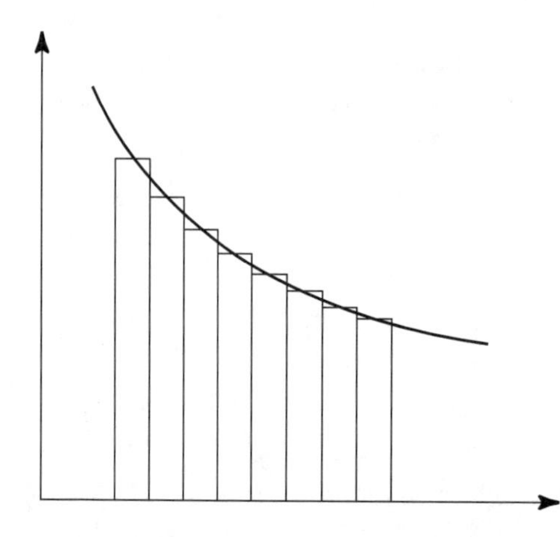

Figure 12-6 Determination of area under curve by the Riemann sum approximation. (Rectangles are chosen so that the triangular areas above the curve approximately equal the triangular areas below the curve. Obviously, the smaller the rectangles, the closer the match between area of the rectangles and the actual area under the curve.)

Many commercial instruments that display data graphically, such as magnetic resonance spectrometers, have built-in electronic integrators that will automatically give the areas under curves. As more and more commercial instrumentation becomes computer interfaced, the task of analyzing the data produced by the instrumentation will become easier.

PROBLEMS

1. Determine the probability of throwing a 2, 3, 4, 5, 6, 7, and 11 with a pair of honest dice.

2. From the following set of data,

Measurement	*Frequency*
5.61	2
5.62	11
5.63	18
5.64	30
5.65	35
5.66	27
5.67	14
5.68	9
5.69	4

determine:

(a) arithmetic mean

(b) standard error of a single measurement

(c) standard error of mean

(d) probable error of a single measurement

(e) probable error of mean

(f) average error

3. The volume of a cylindrical capillary tube is given by the expression $V = \pi r^2 h$, where r is the radius of the capillary tube and h is the height. If the radius of the capillary is found to be 0.030 cm with a probable error of ± 0.002 cm and the height of the capillary is found to be 4.0 cm with a probable error of ± 0.1 cm, what is the volume of the capillary tube and its probable error. What measurement must be made to a higher precision to decrease the probable error in the volume?

4. A student determines the volume of a pycnometer by filling the pycnometer with water and determining its mass. If the mass of the pycnometer plus water is 45.3218 g with a probable error of ± 0.0005 g, the mass of the pycnometer is 25.1011 g with a probable error of ± 0.0005 g, and the density of water at 25°C is 0.997044 g/cc, what is the volume of the pycnometer and its probable error? (Assume the error in the density of water to be negligible.)

5. The molar mass of a vapor is determined by filling a bulb of known volume with the vapor at a known temperature and pressure and measuring its mass. This method is known as the Dumas method. If the vapor is assumed to be an ideal gas, then, from the ideal gas law,

$$M = \frac{mRT}{PV}$$

where M is the molar mass, m is the mass of the vapor, R is the gas constant, T is the absolute temperature, P is the pressure, and V is the volume. Given that

$$m = 1.0339 \pm 0.0007 \text{ g}$$

$$T = 274.0 \pm 0.5 \text{ K}$$

$$P = 1.036 \pm 0.001 \text{ atm}$$

$$V = 0.1993 \pm 0.0001 \text{ liters}$$

$$R = 0.08205 \text{ l} \cdot \text{atm} \cdot \text{mol}^{-1} \cdot \text{K}^{-1} \text{ (no probable error)}$$

determine the error in the molar mass.

6. On millimeter graph paper (using an expanded scale), plot the curve $y = x^3 - 3x^2 + x + 1$ from $x = -1$ to $x = +1$. Using the method of chords, find the slope of the curve at $x = 0$. Compare the slope found by this method to that found by differentiation.

7. On uniform graph paper (using an expanded scale), plot the curve $y = \frac{1}{2}x^2$ from $x = 0$ to $x = 4$. Determine the area under the curve from $x = 1$ to $x = 3$ by cutting out the area and determining its mass on an analytical balance. Next, determine the area by breaking up the area into small rectangles and determining the total area of the rectangles. Compare the areas found by these two methods to the actual area found by integration.

8. Prepare a graph of the data given in Problem 11-10 and determine the change in enthalpy of silver by graphical integration.

9. Prepare a graph of the data given in Problem 11-12 and determine the change in the entropy by graphical integration.

Appendices

I TABLE OF PHYSICAL CONSTANTS

Constant	Symbol	Value (SI Units)
Avogadro's number	N_0	6.022169×10^{23} mol^{-1}
Boltzmann constant	k	1.380662×10^{-23} J · K^{-1}
Electron rest mass	m_e	9.109558×10^{-31} kg
Electron charge	e	1.602191×10^{-19} C
Faraday constant	$\mathcal{F}$	9.648670×10^{4} C · mol^{-1}
Gas constant	R	8.31434×10^{0} J · mol^{-1} · K^{-1}
Permittivity of vacuum	ε_0	8.854188×10^{-12} C^2 · J^{-1} · m
Planck's constant	h	6.626196×10^{-34} J · s
Proton rest mass	m_p	1.672614×10^{-27} kg
Rydberg constant	R_H	1.096776×10^{7} m^{-1}
Speed of light in a vacuum	c	2.997925×10^{8} m · s^{-1}

II TABLE OF INTEGRALS

A. Indefinite Integrals[1]

Algebraic Functions

1. $$\int x^p\,dx = \frac{1}{p+1}x^{p+1} + C, \quad \text{if } p \neq -1$$

2. $$\int \frac{dx}{x} = \ln|x| + C$$

Algebraic Functions of x and ax + b

3. $$\int (ax+b)^p\,dx = \frac{(ax+b)^{p+1}}{a(p+1)} + C, \quad \text{if } p \neq -1$$

4. $$\int \frac{dx}{ax+b} = \frac{1}{a}\ln|ax+b| + C$$

5. $$\int \frac{x\,dx}{ax+b} = \frac{1}{a^2}[ax - b\ln|ax+b|] + C$$

6. $$\int \frac{x\,dx}{(ax+b)^2} = \frac{1}{a^2}\left(\ln|ax+b| + \frac{b}{ax+b}\right) + C$$

[1]From E. J. Cogan and R. Z. Norman, *Handbook of Calculus, Difference, and Differential Equations*, Prentice-Hall, Inc., Englewood Cliffs, New Jersey, 1958, pp. 185–201.

7. $\displaystyle\int \frac{x\,dx}{(ax+b)^3} = \frac{1}{a^2}\left(\frac{b}{2(ax+b)^2} - \frac{1}{ax+b}\right) + C$

8. $\displaystyle\int x(ax+b)^p\,dx = \frac{1}{a^2(p+2)}(ax+b)^{p+2} - \frac{b}{a^2(p+1)}(ax+b)^{p+1} + C, \quad \text{if } p \neq -1,\ p \neq -2$

9. $\displaystyle\int x^m(ax+b)^n\,dx = \frac{1}{m+n+1}[x^{m+1}(ax+b)^n + nb\int x^m(ax+b)^{n-1}\,dx], \quad \text{if } m+n+1 \neq 0$

See (1).

10. $\displaystyle x^m(ax+b)^n\,dx = \frac{1}{a(m+n+1)}[x^m(ax+b)^{n+1} - mb\int x^{m-1}(ax+b)^n\,dx], \quad \text{if } m+n+1 \neq 0$

See (3).

11. $\displaystyle\int \frac{x^2\,dx}{ax+b} = \frac{1}{a^3}\left(\frac{1}{2}(ax+b)^2 - 2b(ax+b) + b^2 \,\mathbf{ln}\, |ax+b|\right) + C$

12. $\displaystyle\int \frac{x^2\,dx}{(ax+b)^2} = \frac{1}{a^3}\left(ax+b - 2b\,\mathbf{ln}\,|ax+b| - \frac{b^2}{ax+b}\right) + C$

13. $\displaystyle\int \sqrt{ax+b}\,dx = \frac{2}{3a}\sqrt{(ax+b)^3} + C$

14. $\displaystyle\int \frac{dx}{\sqrt{ax+b}} = \frac{2\sqrt{ax+b}}{a} + C$

15. $\displaystyle x\sqrt{ax+b}\,dx = \frac{2(3ax-2b)\sqrt{(ax+b)^3}}{15a^2} + C$

16. $\displaystyle\int \frac{\sqrt{ax+b}}{x}\,dx = 2\sqrt{ax+b} + b\int \frac{dx}{x\sqrt{ax+b}}$

See (17) if $b > 0$; (18) if $b < 0$.

17. $\displaystyle\int \frac{dx}{x\sqrt{ax+b}} = \frac{1}{\sqrt{b}}\,\mathbf{ln}\left|\frac{\sqrt{ax+b}-\sqrt{b}}{\sqrt{ax+b}+\sqrt{b}}\right| + C$

$\displaystyle = -\frac{1}{\sqrt{b}}\,\mathbf{tanh}^{-1}\sqrt{\frac{ax+b}{b}} + C, \quad \text{if } b > 0$

18. $\displaystyle\int \frac{dx}{x\sqrt{ax-b}} = \frac{2}{\sqrt{b}}\,\mathbf{arctan}\sqrt{\frac{ax-b}{b}} + C, \quad \text{if b} > 0$

19. $\displaystyle x^2\sqrt{ax+b}\,dx = \frac{2}{105a^3}\sqrt{(ax+b)^3}(15a^2x^2 - 12abx + 8b^2) + C$

20. $\displaystyle\int \frac{x^n dx}{\sqrt{ax+b}} = \frac{2}{a(2n+1)}\left[x^n\sqrt{ax+b} - nb\int \frac{x^{n-1}dx}{\sqrt{ax+b}}\right]$

21. $$\int \frac{dx}{x^n\sqrt{ax+b}} = -\frac{\sqrt{ax+b}}{(n-1)bx^{n-1}} - \frac{(2n-3)a}{(2n-2)b}\int \frac{dx}{x^{n-1}\sqrt{ax+b}}$$

See (14).

Algebraic Functions of x and ax²+b

22. $$\int \frac{dx}{ax^2+b} = \frac{1}{\sqrt{ab}}\ \mathbf{arctan}\ \frac{x\sqrt{ab}}{b} + C, \quad \text{if } ab > 0$$

23. $$\int \frac{dx}{ax^2+b} = \frac{1}{2\sqrt{-ab}}\ \mathbf{ln}\left|\frac{b+x\sqrt{-ab}}{b-x\sqrt{-ab}}\right| + C$$

$$= \frac{1}{\sqrt{-ab}}\ \mathbf{tanh}^{-1}\frac{x\sqrt{-ab}}{b} + C, \quad \text{if } ab < 0$$

24. $$\int \frac{dx}{ax^2-b} = \frac{1}{2\sqrt{ab}}\ \mathbf{ln}\left|\frac{x\sqrt{a}-\sqrt{b}}{x\sqrt{a}+\sqrt{b}}\right| + C, \quad \text{if } a>0, b>0$$

25. $$\int \frac{dx}{-ax^2+b} = \frac{1}{2\sqrt{ab}}\ \mathbf{ln}\left|\frac{\sqrt{b}+x\sqrt{a}}{\sqrt{b}-x\sqrt{a}}\right| + C, \quad \text{if } a>0, b>0$$

26. $$\int \frac{dx}{(ax^2+b)^{m+1}} = \frac{1}{2mb}\left[\frac{x}{(ax^2+b)^m} + (2m-1)\int \frac{dx}{(ax^2+b)^m}\right]$$

See (23) or (24) if $b > 0$; (23) if $b < 0$.

27. $$\int x(ax^2+b)^p\,dx = \frac{1}{2a}\frac{(ax^2+b)^{p+1}}{p+1} + C, \quad \text{if } p \neq -1$$

28. $$\int \frac{x\,dx}{ax^2+b} = \frac{1}{2a}\ \mathbf{ln}\,|ax^2+b| + C$$

29. $$\int \frac{x\,dx}{(ax^2+b)^{p+1}} = \frac{-1}{2pa(ax^2+b)^p} + C, \quad \text{if } p > 0$$

30. $$\int \frac{dx}{x(ax^n+b)} = \frac{1}{bn}\ \mathbf{ln}\left|\frac{x^n}{ax^n+b}\right| + C$$

31. $$\int x^p(ax^n+b)^m\,dx = \frac{1}{nm+p+1}[x^{p+1}(ax^n+b)^m + bnm\int x^p(ax^n+b)^{m-1}\,dx],$$

if $m > 0$ and $nm+p+1 \neq 0$

See (1).

32. $$\int \sqrt{ax^2+b}\,dx = \frac{x}{2}\sqrt{ax^2+b} + \frac{b}{2\sqrt{a}}\ \mathbf{ln}\left|x\sqrt{a}+\sqrt{ax^2+b}\right| + C, \quad \text{if } a > 0$$

33. $$\int \sqrt{ax^2+b}\,dx = \frac{x}{2}\sqrt{ax^2+b} + \frac{b}{2\sqrt{-a}}\ \mathbf{arcsin}\left(x\sqrt{\frac{-a}{b}}\right) + C, \quad \text{if } a < 0$$

Algebraic Functions of x and $ax^2 + bx + c$

$$X = ax^2 + bx + c, \quad q = 4ac - b^2$$

34. $\displaystyle\int \frac{dx}{ax^2 + bx + c} = \frac{2}{\sqrt{q}} \textbf{ arctan } \frac{2ax + b}{\sqrt{q}} + C, \quad \text{if } q > 0$

35. $\displaystyle\int \frac{dx}{ax^2 + bx + c} = \frac{2}{\sqrt{-q}} \textbf{ ln } \left| \frac{2ax + b - \sqrt{-q}}{2ax + b + \sqrt{-q}} \right| + C$

$\displaystyle\qquad = \frac{-2}{\sqrt{-q}} \textbf{ tanh}^{-1} \frac{2ax + b}{\sqrt{-q}} + C, \quad \text{if } q < 0$

36. $\displaystyle\int \frac{dx}{(ax^2 + bx + c)^{n+1}} = \frac{2ax + b}{nqX^n} + \frac{2(2n - 1)a}{nq} \int \frac{dx}{X^n}$

See (34) if $q > 0$; (35) if $q < 0$.

37. $\displaystyle\int \sqrt{ax^2 + bx + c}\, dx = \frac{2ax + b}{4a} \sqrt{X} + \frac{q}{8a} \int \frac{dx}{\sqrt{X}}$

See (38) if $a > 0$; (39) if $a < 0$.

38. $\displaystyle\int \frac{dx}{\sqrt{ax^2 + bx + c}} = \frac{1}{\sqrt{a}} \textbf{ ln } \left| X + x\sqrt{a} + \frac{b}{2\sqrt{a}} \right| + C, \quad \text{if } a > 0$

39. $\displaystyle\int \frac{dx}{\sqrt{ax^2 + bx + c}} = \frac{1}{\sqrt{-a}} \textbf{ arcsin } \frac{-2ax - b}{\sqrt{-q}} + C, \quad \text{if } a < 0$

40. $\displaystyle\int \frac{x\, dx}{ax^2 + bx + c} = \frac{1}{2a} \left[\textbf{ln}|X| - b \int \frac{dx}{X} \right]$

See (32) if $q > 0$; (33) if $q < 0$.

Algebraic Functions of x and $x^2 \pm a^2$

41. $\displaystyle\int \frac{dx}{x^2 + a^2} = \frac{1}{a} \textbf{ arctan } \frac{x}{a} + C$

42. $\displaystyle\int \frac{dx}{x^2 - a^2} = \frac{1}{2a} \textbf{ ln } \left| \frac{x - a}{x + a} \right| + C = -\frac{1}{a} \textbf{ coth}^{-1} \frac{x}{a} + C$

43. $\displaystyle\int \sqrt{x^2 \pm a^2}\, dx = \frac{1}{2} \left[x\sqrt{x^2 \pm a^2} \pm a^2 \textbf{ ln } \left| x + \sqrt{x^2 \pm a^2} \right| \right] + C$

44. $\displaystyle\int \frac{dx}{\sqrt{x^2 \pm a^2}} = \textbf{ln } \left| x + \sqrt{x^2 \pm a^2} \right| + C$

45. $\displaystyle\int x\sqrt{x^2 \pm a^2}\, dx = \frac{1}{3} \sqrt{(x^2 \pm a^2)^3} + C$

46. $\displaystyle\int \frac{\sqrt{x^2 + a^2}}{x}\, dx = \sqrt{x^2 + a^2} - a \textbf{ ln } \left| \frac{a + \sqrt{x^2 + a^2}}{x} \right| + C$

47. $\displaystyle\int \frac{\sqrt{x^2 - a^2}}{x}\, dx = \sqrt{x^2 - a^2} - a \textbf{ arcsec } \frac{x}{a} + C$

48. $$\int \frac{x\,dx}{\sqrt{x^2 \pm a^2}} = \sqrt{x^2 \pm a^2} + C$$

49. $$\int \frac{dx}{x\sqrt{x^2 + a^2}} = -\frac{1}{a} \ln \left| \frac{\sqrt{x^2 + a^2} + a}{x} \right| + C = -\frac{1}{a} \sinh^{-1} \frac{a}{x} + C$$

50. $$\int \frac{dx}{x\sqrt{x^2 - a^2}} = \frac{1}{a} \operatorname{arcsec} \frac{x}{a} + C = \frac{1}{a} \arccos \frac{a}{x} + C$$

51. $$\int \sqrt{(x^2 \pm a^2)^3}\,dx = \frac{1}{8}\left[2x\sqrt{(x^2 \pm a^2)^3} \pm 3a^2 x\sqrt{x^2 \pm a^2} + 3a^4 \ln \left| x - \sqrt{x^2 \pm a^2} \right| \right] + C$$

52. $$\int \frac{dx}{\sqrt{(x^2 \pm a^2)^3}} = \frac{\pm x}{a^2\sqrt{x^2 \pm a^2}} + C$$

53. $$\int x\sqrt{(x^2 \pm a^2)^3}\,dx = \frac{1}{5}\sqrt{(x^2 \pm a^2)^5} + C$$

54. $$\int \frac{x\,dx}{(x^2 \pm a^2)^3} = \frac{-1}{\sqrt{x^2 \pm a^2}} + C$$

Algebraic Functions of x and $a^2 - x^2$

55. $$\int \frac{dx}{a^2 - x^2} = \frac{1}{2a} \ln \left| \frac{a + x}{a - x} \right| + C = \frac{1}{a} \tanh^{-1} \frac{x}{a} + C$$

56. $$\int \sqrt{a^2 - x^2}\,dx = \frac{1}{2}\left[x\sqrt{a^2 - x^2} + a^2 \arcsin \frac{x}{a}\right] + C$$

57. $$\int \frac{dx}{\sqrt{a^2 - x^2}} = \arcsin \frac{x}{a} + C = -\arccos \frac{x}{a} + C$$

58. $$\int x\sqrt{a^2 - x^2}\,dx = -\frac{1}{3}\sqrt{(a^2 - x^2)^3} + C$$

59. $$\int \frac{\sqrt{a^2 - x^2}}{x}\,dx = \sqrt{a^2 - x^2} - a \ln \left| \frac{a + \sqrt{a^2 - x^2}}{x} \right| + C$$

60. $$\int \frac{x\,dx}{\sqrt{a^2 - x^2}} = -\sqrt{a^2 - x^2} + C$$

61. $$\int \frac{dx}{x\sqrt{a^2 - x^2}} = -\frac{1}{a} \ln \left| \frac{a + \sqrt{a^2 - x^2}}{x} \right| + C = -\frac{1}{a} \cosh^{-1} \frac{a}{x} + C$$

62. $$\int \sqrt{(a^2 - x^2)^3}\,dx = \frac{1}{8}\left[2x\sqrt{(a^2 - x^2)^3} + 3a^2 x\sqrt{a^2 - x^2} + 3a^4 \arcsin \frac{x}{a}\right] + C$$

63. $$\int \frac{dx}{\sqrt{(a^2 - x^2)^3}} = \frac{x}{a^2\sqrt{a^2 - x^2}} + C$$

64. $\int x\sqrt{(a^2-x^2)^3}\,dx = -\frac{1}{5}\sqrt{(a^2-x^2)^5}+C$

65. $\int \frac{x\,dx}{\sqrt{(a^2-x^2)^3}} = \frac{1}{\sqrt{a^2-x^2}}+C$

Other Algebraic Functions

66. $\int \sqrt{2ax-x^2}\,dx = \frac{1}{2}\left[(x-a)\sqrt{2ax-x^2}+a^2\,\mathbf{arcsin}\,\frac{x-a}{a}\right]+C$

67. $\int \frac{dx}{\sqrt{2ax-x^2}} = \mathbf{arccos}\,\frac{a-x}{a}+C$

68. $\int \sqrt{\frac{1+x}{1-x}}\,dx = \mathbf{arcsin}\,x-\sqrt{1-x^2}+C$

Trigonometric Functions

69. $\int \mathbf{sin}\,x\,dx = -\mathbf{cos}\,x+C$

70. $\int \mathbf{sin}\,(ax+b)\,dx = -\frac{1}{a}\,\mathbf{cos}\,(ax+b)+C$

71. $\int \mathbf{sin}^2\,(ax+b)\,dx = \frac{x}{2}-\frac{1}{2a}\,\mathbf{cos}\,(ax+b)\,\mathbf{sin}\,(ax+b)+C$

$$= \frac{x}{2}-\frac{\mathbf{sin}\,2(ax+b)}{4a}+C$$

72. $\int \mathbf{sin}^3\,(ax+b)\,dx = -\frac{1}{3a}\,\mathbf{cos}\,(ax+b)[\mathbf{sin}^2\,(ax+b)+2]+C$

73. $\int \mathbf{sin}^4\,(ax+b)\,dx = \frac{3x}{8}-\frac{3\,\mathbf{sin}\,2(ax+b)}{16a}-\frac{\mathbf{sin}^3\,(ax+b)\,\mathbf{cos}\,(ax+b)}{4a}+C$

74. $\int \mathbf{sin}^n\,(ax+b)\,dx = -\frac{1}{an}\left[\mathbf{sin}^{n-1}\,(ax+b)\,\mathbf{cos}\,(ax+b)\right.$

$$\left.-\,a(n-1)\int \mathbf{sin}^{n-2}\,(ax+b)\,dx\right]$$

See (73) if n is even; (72) if n is odd.

75. $\int x\,\mathbf{sin}\,(ax+b)\,dx = \frac{1}{a^2}\,\mathbf{sin}\,(ax+b)-\frac{x}{a}\,\mathbf{cos}\,(ax+b)+C$

76. $\int x\,\mathbf{sin}^2\,(ax+b)\,dx = \frac{x^2}{4}-\frac{x\,\mathbf{sin}\,2(ax+b)}{4a}-\frac{\mathbf{cos}\,2(ax+b)}{8a^2}+C$

77. $$\int x^2 \sin^2(ax+b)\,dx = \frac{x^3}{6} - \left(\frac{x^2}{4a} - \frac{1}{8a^3}\right)\sin 2(ax+b) - \frac{x\cos 2(ax+b)}{4a^2} + C$$

78. $$\int x^n \sin(ax+b)\,dx = -\frac{1}{a}\left[x^n \cos(ax+b) - n\int x^{n-1}\cos(ax+b)\,dx\right]$$

See (91) if n is even; (75) if n is odd.

79. $$\int \frac{\sin ax\,dx}{x} = ax - \frac{(ax)^3}{3(3!)} + \frac{(ax)^5}{5(5!)} - + \cdots + C$$

80. $$\int \frac{\sin(ax+b)\,dx}{x^n} = \frac{-1}{n-1}\frac{\sin(ax+b)}{x^{n-1}} + \frac{a}{n-1}\int \frac{\cos(xa+b)\,dx}{x^{n-1}}, \quad \text{if } n > 1$$

See (70) if n is even; (86) if n is odd.

81. $$\int \frac{dx}{1 \pm \sin(ax+b)} = \mp\frac{1}{a}\tan\left(\frac{\pi}{4} \mp \frac{ax+b}{2}\right) + C$$

82. $$\int \sqrt{1+\sin x}\,dx = \pm 2\left(\sin\frac{x}{2} - \cos\frac{x}{2}\right) + C$$

Use + sign if $(8n-1)\frac{\pi}{2} < x \le (8n+3)\frac{\pi}{2}$ for some integer n; otherwise use − sign.

83. $$\int \sqrt{1-\sin x}\,dx = \pm 2\left[\sin\frac{x}{2} + \cos\frac{x}{2}\right] + C$$

Use + sign if $(8n-3)\frac{\pi}{2} < x \le (8n+1)\frac{\pi}{2}$ for some integer n; otherwise use − sign.

84. $$\int \sin ax \sin bx\,dx = \frac{\sin(a-b)x}{2(a-b)} - \frac{\sin(a+b)x}{2(a+b)} + C, \quad \text{if } a^2 \ne b^2$$

85. $$\int \cos x\,dx = \sin x + C$$

86. $$\int \cos(ax+b)\,dx = \frac{1}{a}\sin(ax+b) + C$$

87. $$\int \cos^2(ax+b)\,dx = \frac{x}{2} + \frac{1}{2a}\cos(ax+b)\sin(ax+b) + C = \frac{x}{2} + \frac{\sin 2(ax+b)}{4a} + C$$

88. $$\int \cos^3(ax+b)\,dx = \frac{1}{a}\sin(ax+b) - \frac{1}{3a}\sin^3(ax+b) + C$$

89. $$\int \cos^4(ax+b)\,dx = \frac{3x}{8} + \frac{3\sin 2(ax+b)}{16a} + \frac{\cos^3(ax+b)\sin(ax+b)}{4a} + C$$

90. $\displaystyle\int \mathbf{cos}^n\,(ax+b)\,dx = \frac{1}{an}\left[\mathbf{cos}^{n-1}\,(ax+b)\,\mathbf{sin}\,(ax+b) + a(n-1)\int \mathbf{cos}^{n-2}\,(ax+b)\,dx\right]$

See (88) if n is odd; (89) if n is even.

91. $\displaystyle\int x\,\mathbf{cos}\,(ax+b)\,dx = \frac{1}{a^2}\,\mathbf{cos}\,(ax+b) + \frac{x}{a}\,\mathbf{sin}\,(ax+b) + C$

92. $\displaystyle\int x\,\mathbf{cos}^2\,(ax+b)\,dx = \frac{x^2}{4} + \frac{x\,\mathbf{sin}\,2(ax+b)}{4a} + \frac{\mathbf{cos}\,2(ax+b)}{8a^2} + C$

93. $\displaystyle\int x^2\,\mathbf{cos}^2\,(ax+b)\,dx = \frac{x^3}{6} + \left(\frac{x^2}{4a} - \frac{1}{8a^3}\right)\mathbf{sin}\,2(ax+b) + \frac{x\,\mathbf{cos}\,2(ax+b)}{4a^2} + C$

94. $\displaystyle\int x^n\,\mathbf{cos}\,(ax+b)\,dx = \frac{1}{a}\left[x^n\,\mathbf{sin}\,(ax+b) - n\int x^{n-1}\,\mathbf{sin}\,(ax+b)\,dx\right] + C$

See (75) if n is even; (91) if n is odd.

95. $\displaystyle\int \frac{\mathbf{cos}\,ax\,dx}{x} = \mathbf{ln}\,|ax| - \frac{(ax)^2}{2(2!)} + \frac{(ax)^4}{4(4!)} - + \cdots + C$

96. $\displaystyle\int \frac{\mathbf{cos}\,(ax+b)\,dx}{x^n} = \frac{-1}{n-1}\,\frac{\mathbf{cos}\,(ax+b)}{x^{n-1}} - \frac{a}{n-1}\int \frac{\mathbf{sin}\,(ax+b)\,dx}{x^{n-1}}, \quad \text{if } n > 1$

See (86) if n is even; (70) if n is odd.

97. $\displaystyle\int \frac{dx}{1+\mathbf{cos}\,(ax+b)} = \frac{1}{a}\,\mathbf{tan}\left(\frac{ax+b}{2}\right) + C$

98. $\displaystyle\int \frac{dx}{1-\mathbf{cos}\,(ax+b)} = -\frac{1}{a}\,\mathbf{cot}\left(\frac{ax+b}{2}\right) + C$

99. $\displaystyle\int \sqrt{1+\mathbf{cos}\,x}\,dx = \pm 2\sqrt{2}\,\mathbf{sin}\,\frac{x}{2} + C$

Use + sign if $(4n-1)\pi < x \le (4n+1)\pi$ for some integer n; otherwise use − sign.

100. $\displaystyle\int \sqrt{1-\mathbf{cos}\,x}\,dx = \pm 2\sqrt{2}\,\mathbf{cos}\,\frac{x}{2} + C$

Use + sign if $(4n-2)\pi < x \le 4n\pi$ for some integer n; otherwise use − sign.

101. $\displaystyle\int \mathbf{cos}\,ax\,\mathbf{cos}\,bx\,dx = \frac{\mathbf{sin}\,(a-b)x}{2(a-b)} + \frac{\mathbf{sin}\,(a+b)x}{2(a+b)} + C, \quad \text{if } a^2 \ne b^2$

102. $\displaystyle\int \mathbf{sin}\,ax\,\mathbf{cos}\,bx\,dx = -\frac{\mathbf{cos}\,(a-b)x}{2(a-b)} - \frac{\mathbf{cos}\,(a+b)x}{2(a+b)} + C, \quad \text{if } a^2 \ne b^2$

103. $\displaystyle\int \mathbf{sin}\,(ax+b)\,\mathbf{cos}\,(ax+b)\,dx = \frac{1}{2a}\,\mathbf{sin}^2\,(ax+b) + C$

104. $$\int \sin^p (ax+b) \cos (ax+b)\, dx = \frac{1}{a(p+1)} \sin^{p+1} (ax+b) + C, \quad \text{if } p \neq -1$$

105. $$\int \sin (ax+b) \cos^p (ax+b)\, dx = -\frac{1}{a(p+1)} \cos^{p+1} (ax+b) + C,$$

if $p \neq -1$

106. $$\int \sin^2 (ax+b) \cos^2 (ax+b)\, dx = -\frac{1}{32a} \sin 4(ax+b) + \frac{x}{8} + C$$

107. $$\int \sin^m (ax+b) \cos^n (ax+b)\, dx$$
$$= -\frac{1}{a(m+n)} \left[\sin^{m-1} (ax+b) \cos^{n+1} (ax+b) - (m-1)a \int \sin^{m-2} (ax+b) \cos^n (ax+b)\, dx \right], \quad \text{if } m > 0,\ m+n \neq 0$$

See (90) if m is even; (105) if m is odd.

108. $$\int \sin^m (ax+b) \cos^n (ax+b)\, dx$$
$$= \frac{1}{a(m+n)} \left[\sin^{m+1} (ax+b) \cos^{n-1} (ax+b) + (n-1)a \int \sin^m (ax+b) \cos^{n-2} (ax+b)\, dx \right], \quad \text{if } n > 0,\ m+n \neq 0$$

See (74) if n is even; (104) if n is odd.

109. $$\int \frac{dx}{\sin (ax+b) \cos (ax+b)} = \frac{1}{a} \ln |\tan (ax+b)| + C$$

110. $$\int \frac{dx}{\sin^n (ax+b) \cos (ax+b)} = \frac{-1}{a(n-1)} \frac{1}{\sin^{n-1} (ax+b)} + \int \frac{dx}{\sin^{n-2} (ax+b) \cos (ax+b)}, \quad \text{if } n > 1$$

See (128) if n is even; (109) if n is odd.

111. $$\int \frac{dx}{\sin (ax+b) \cos^n (ax+b)} = \frac{1}{a(n-1)} \frac{1}{\cos^{n-1} (ax+b)} + \int \frac{dx}{\sin (ax+b) \cos^{n-2} (ax+b)}, \quad \text{if } n > 1$$

See (135) if n is even; (109) if n is odd.

112. $$\int \frac{\sin (ax+b)}{\cos^2 (ax+b)}\, dx = \frac{1}{a \cos (ax+b)} + C = \frac{1}{a} \sec (ax+b) + C$$

113. $$\int \frac{\sin^2 (ax+b)}{\cos (ax+b)}\, dx = -\frac{1}{a} \left[\sin (ax+b) - \ln \left| \tan \left(\frac{ax+b}{2} + \frac{\pi}{4} \right) \right| \right] + C$$

114. $\displaystyle\int \frac{\cos(ax+b)}{\sin^2(ax+b)}\,dx = \frac{-1}{a\sin(ax+b)} + C = -\frac{1}{a}\csc(ax+b) + C$

115. $\displaystyle\int \frac{\cos^2(ax+b)}{\sin(ax+b)}\,dx = \frac{1}{a}\left[\cos(ax+b) + \ln\left|\tan\left(\frac{ax+b}{2}\right)\right|\right] + C$

116. $\displaystyle\int \tan x\,dx = -\ln|\cos x| + C = \ln|\sec x| + C$

117. $\displaystyle\int \tan(ax+b)\,dx = -\frac{1}{a}\ln|\cos(ax+b)| + C$

118. $\displaystyle\int \tan^2(ax+b)\,dx = \frac{1}{a}\tan(ax+b) - x + C$

119. $\displaystyle\int \tan^3(ax+b)\,dx = \frac{1}{2a}[\tan^2(ax+b) + 2\ln|\cos(ax+b)|] + C$

120. $\displaystyle\int \tan^n(ax+b)\,dx = \frac{1}{a(n-1)}\tan^{n-1}(ax+b) - \int \tan^{n-2}(ax+b)\,dx, \quad n \geq 2$

See (118) if n is even; (119) if n is odd.

121. $\displaystyle\int \cot x\,dx = \ln|\sin x| + C = -\ln|\csc x| + C$

122. $\displaystyle\int \cot(ax+b)\,dx = \frac{1}{a}\ln|\sin(ax+b)| + C$

123. $\displaystyle\int \cot^2(ax+b)\,dx = -\frac{1}{a}\cot(ax+b) - x + C$

124. $\displaystyle\int \cot^3(ax+b)\,dx = -\frac{1}{2a}[\cot^2(ax+b) + 2\ln|\sin(ax+b)|] + C$

125. $\displaystyle\int \cot^n(ax+b)\,dx = -\frac{1}{a(n-1)}\cot^{n-1}(ax+b) - \int \cot^{n-2}(ax+b)\,dx, \quad n \geq 2$

See (123) if n is even; (124) if n is odd.

126. $\displaystyle\int \sec x\,dx = \ln\left|\tan\left(\frac{x}{2} + \frac{\pi}{4}\right)\right| + C = \ln|\sec x + \tan x| + C$

127. $\displaystyle\int \sec^2 x\,dx = \tan x + C$

128. $\displaystyle\int \sec(ax+b)\,dx = \frac{1}{a}\ln\left|\tan\left(\frac{ax+b}{2} + \frac{\pi}{4}\right)\right| + C$

129. $\displaystyle\int \sec^2(ax+b)\,dx = \frac{1}{a}\tan(ax+b) + C$

130. $$\int \sec^3 (ax+b)\,dx = \frac{1}{2a}\left[\sec (ax+b) \tan (ax+b) + \ln \left|\tan \left(\frac{ax+b}{2}+\frac{\pi}{4}\right)\right|\right] + C$$

131. $$\int \sec^n (ax+b)\,dx = \frac{1}{a(n-1)}\frac{\sin (ax+b)}{\cos^{n-1} (ax+b)} + \frac{n-2}{n-1}\int \sec^{n-2} (ax+b)\,dx, \quad n \geq 2$$

See (129) if n is even; (130) if n is odd.

132. $$\int \sec x \tan x\,dx = \sec x + C$$

133. $$\int \csc x\,dx = \ln \left|\tan \left(\frac{x}{2}\right)\right| + C = \ln |\csc x - \cot x| + C$$

134. $$\int \csc^2 x\,dx = -\cot x + C$$

135. $$\int \csc (ax+b)\,dx = \frac{1}{a}\ln \left|\tan \left(\frac{ax+b}{2}\right)\right| + C$$

136. $$\int \csc^2 (ax+b)\,dx = -\frac{1}{a}\cot (ax+b) + C$$

137. $$\int \csc^3 (ax+b)\,dx = \frac{1}{2a}\left[-\csc (ax+b) \cot (ax+b) + \ln \left|\tan \left(\frac{ax+b}{2}\right)\right|\right] + C$$

138. $$\int \csc^n (ax+b)\,dx = \frac{-1}{a(n-1)}\frac{\cos (ax+b)}{\sin^{n-1} (ax+b)} + \frac{n-2}{n-1}\int \csc^{n-2} (ax+b)\,dx, \quad n \geq 2$$

See (136) if n is even; (137) if n is odd.

139. $$\int \csc x \cot x\,dx = -\csc x + C$$

140. $$\int \arcsin \frac{x}{a}\,dx = x \arcsin \frac{x}{a} + \sqrt{a^2-x^2} + C$$

141. $$\int (\arcsin ax)^2\,dx = x (\arcsin ax)^2 - 2x + \frac{2}{a}\sqrt{1-a^2x^2} \arcsin ax + C$$

142. $$\int x \arcsin ax\,dx = \frac{1}{4a^2}\left[(2a^2x^2-1) \arcsin ax + ax\sqrt{1-a^2x^2}\right] + C$$

143. $\int \frac{\textbf{arcsin}\, ax}{x^2}\, dx = a\, \textbf{ln} \left| \frac{1 - \sqrt{1 - a^2x^2}}{ax} \right| - \frac{\textbf{arcsin}\, ax}{x} + C$

144. $\int \textbf{arccos}\, \frac{x}{a}\, dx = x\, \textbf{arccos}\, \frac{x}{a} - \sqrt{a^2 - x^2} + C$

145. $\int (\textbf{arccos}\, ax)^2\, dx = x\, (\textbf{arccos}\, ax)^2 - 2x - \frac{2}{a}\sqrt{1 - a^2x^2}\, \textbf{arccos}\, ax + C$

146. $\int \textbf{arctan}\, \frac{x}{a}\, dx = x\, \textbf{arctan}\, \frac{x}{a} - \frac{a}{2}\, \textbf{ln}\, (a^2 + x^2) + C$

147. $\int \textbf{arccot}\, \frac{x}{a}\, dx = x\, \textbf{arccot}\, \frac{x}{a} + \frac{a}{2}\, \textbf{ln}\, (a^2 + x^2) + C$

148. $\int \textbf{arcsec}\, \frac{x}{a}\, dx = x\, \textbf{arcsec}\, \frac{x}{a} - a\, \textbf{ln}\, \left| x + \sqrt{x^2 - a^2} \right| + C$

149. $\int \textbf{arccsc}\, \frac{x}{a}\, dx = x\, \textbf{arccsc}\, \frac{x}{a} + a\, \textbf{ln}\, \left| x + \sqrt{x^2 - a^2} \right| + C$

Logarithmic Functions

150. $\int \textbf{ln}\, |x|\, dx = x\, \textbf{ln}\, |x| - x + C$

151. $\int \textbf{log}_a\, |x|\, dx = x\, \textbf{log}_a\, |x| - \frac{x}{\textbf{ln}\, a} + C, \quad \text{if } a \neq 1, a > 0$

152. $\int \textbf{ln}\, |ax + b|\, dx = \frac{ax + b}{a}\, \textbf{ln}\, |ax + b| - x + C$

153. $\int (\textbf{ln}\, |x|)^2\, dx = x\, (\textbf{ln}\, |x|)^2 - 2x\, \textbf{ln}\, |x| + 2x + C$

154. $\int (\textbf{ln}\, |ax + b|)^n\, dx = \frac{ax + b}{a}(\textbf{ln}\, |ax + b|)^n - n \int (\textbf{ln}\, |ax + b|)^{n-1}\, dx$

See (152).

155. $\int x\, \textbf{ln}\, |x|\, dx = \frac{x^2}{2}\, \textbf{ln}\, |x| - \frac{x^2}{4} + C$

156. $\int \frac{dx}{x\, \textbf{ln}\, |x|} = \textbf{ln}\, \left| \textbf{ln}\, |x| \right| + C$

157. $\int x^p\, \textbf{ln}\, |x|\, dx = x^{p+1} \left[\frac{\textbf{ln}\, |x|}{p + 1} - \frac{1}{(p + 1)^2} \right] + C, \quad \text{if } p \neq -1$

158. $\int \frac{(\textbf{ln}\, |x|)^p}{x}\, dx = \frac{1}{p + 1}(\textbf{ln}\, |x|)^{p+1} + C, \quad \text{if } p \neq -1$

159. $\int \textbf{sin}\, (\textbf{ln}\, |x|)\, dx = \frac{x}{2}[\textbf{sin}\, (\textbf{ln}\, |x|) - \textbf{cos}\, (\textbf{ln}\, |x|)] + C$

160. $\int \textbf{cos}\, (\textbf{ln}\, |x|)\, dx = \frac{x}{2}[\textbf{sin}\, (\textbf{ln}\, |x|) + \textbf{cos}\, (\textbf{ln}\, |x|)] + C$

Exponential Functions

161. $\int \mathbf{e}^x \, dx = \mathbf{e}^x + C$

162. $\int \mathbf{e}^{ax} \, dx = \frac{1}{a} \, \mathbf{e}^{ax} + C$

163. $\int x \, \mathbf{e}^{ax} \, dx = \frac{1}{a^2} \, \mathbf{e}^{ax} \, (ax - 1) + C$

164. $\int x^m \, \mathbf{e}^{ax} \, dx = \frac{1}{a} x^m \, \mathbf{e}^{ax} - \frac{m}{a} \int x^{m-1} \, \mathbf{e}^{ax} \, dx, \quad m \geq 2$

See (163).

165. $\int \frac{\mathbf{e}^{ax} \, dx}{x} = \mathbf{ln} \, |x| + ax + \frac{(ax)^2}{2(2!)} + \frac{(ax)^3}{3(3!)} + \cdots + C$

166. $\int \mathbf{e}^{ax} \, \mathbf{sin} \, bx \, dx = \frac{\mathbf{e}^{ax} \, (a \, \mathbf{sin} \, bx - b \, \mathbf{cos} \, bx)}{a^2 + b^2} + C$

167. $\int \mathbf{e}^{ax} \, \mathbf{cos} \, bx \, dx = \frac{\mathbf{e}^{ax} \, (a \, \mathbf{cos} \, bx + b \, \mathbf{sin} \, bx)}{a^2 + b^2} + C$

168. $\int \frac{dx}{1 + \mathbf{e}^x} = x - \mathbf{ln} \, |1 + \mathbf{e}^x| + C$

169. $\int \frac{dx}{a \, \mathbf{e}^{px} + b} = \frac{x}{b} - \frac{1}{bp} \, \mathbf{ln} \, |a \, \mathbf{e}^{px} + b| + C, \quad \text{if } b \neq 0, \, p \neq 0$

170. $\int \frac{dx}{a \, \mathbf{e}^{px} + b \, \mathbf{e}^{-px}} = \frac{1}{p\sqrt{ab}} \, \mathbf{arctan} \left(e^{px} \sqrt{\frac{a}{b}} \right) + C, \quad \text{if } ab > 0$

171. $\int \mathbf{e}^{ax} \, \mathbf{ln} \, |bx| \, dx = \frac{1}{a} \, \mathbf{e}^{ax} \, \mathbf{ln} \, |bx| - \frac{1}{a} \int \frac{\mathbf{e}^{ax}}{x} \, dx$

See (165).

172. $\int \mathbf{a}^x \, dx = \frac{\mathbf{a}^x}{\mathbf{ln} \, a} + C, \quad \text{if } a > 0, a \neq 1$

173. $\int \mathbf{a}^{bx} \, dx = \frac{\mathbf{a}^{bx}}{b \, \mathbf{ln} \, a} + C, \quad a > 0, a \neq 1$

174. $\int x \, \mathbf{a}^{bx} \, dx = \frac{x \, \mathbf{a}^{bx}}{b \, \mathbf{ln} \, a} - \frac{\mathbf{a}^{bx}}{b^2 (\mathbf{ln} \, a)^2} + C, \quad \text{if } a > 0, a \neq 1$

Hyperbolic Functions

175. $\int \mathbf{sinh} \, ax \, dx = \frac{1}{a} \, \mathbf{cosh} \, ax + C$

176. $\int \mathbf{sinh}^2 \, ax \, dx = \frac{1}{4a} \, \mathbf{sinh} \, 2ax - \frac{1}{2} x + C$

177. $\int \mathbf{cosh} \, ax \, dx = \frac{1}{a} \, \mathbf{sinh} \, ax + C$

178. $\displaystyle\int \mathbf{cosh}^2\, ax\, dx = \frac{1}{4a}\, \mathbf{sinh}\, 2ax + \frac{1}{2}x + C$

179. $\displaystyle\int \mathbf{tanh}\, ax\, dx = \frac{1}{a}\, \mathbf{ln}\, |\mathbf{cosh}\, ax| + C$

180. $\displaystyle\int \mathbf{tanh}^2\, ax\, dx = x - \frac{1}{a}\, \mathbf{tanh}\, ax + C$

181. $\displaystyle\int \mathbf{coth}\, ax\, dx = \frac{1}{a}\, \mathbf{ln}\, |\mathbf{sinh}\, ax| + C$

182. $\displaystyle\int \mathbf{coth}^2\, ax\, dx = x - \frac{1}{a}\, \mathbf{coth}\, ax + C$

183. $\displaystyle\int \mathbf{sech}\, ax\, dx = \frac{1}{a}\, \mathbf{arctan}\, (\mathbf{sinh}\, ax) + C$

184. $\displaystyle\int \mathbf{sech}^2\, ax\, dx = \frac{1}{a}\, \mathbf{tanh}\, ax + C$

185. $\displaystyle\int \mathbf{csch}\, ax\, dx = -\frac{1}{a}\, \mathbf{ln}\, |\mathbf{coth}\, ax + \mathbf{csch}\, ax| + C = \frac{1}{a}\, \mathbf{ln}\, \left|\mathbf{tanh}\, \frac{ax}{2}\right| + C$

186. $\displaystyle\int \mathbf{csch}^2\, ax\, dx = -\frac{1}{a}\, \mathbf{coth}\, ax + C$

187. $\displaystyle\int \mathbf{sech}\, ax\, \mathbf{tanh}\, ax\, dx = -\frac{1}{a}\, \mathbf{sech}\, ax + C$

188. $\displaystyle\int \mathbf{csch}\, ax\, \mathbf{coth}\, ax\, dx = -\frac{1}{a}\, \mathbf{csch}\, ax + C$

B. Definite Integrals

1. $\displaystyle\int_0^\infty \mathbf{e}^{-ax^2}\, dx = \frac{1}{2}\left(\frac{\pi}{a}\right)^{1/2}$

2. $\displaystyle\int_0^\infty x^2\, \mathbf{e}^{-ax^2}\, dx = \frac{1}{4a}\left(\frac{\pi}{a}\right)^{1/2}$

3. $\displaystyle\int_0^\infty x^{2n}\, \mathbf{e}^{-ax^2}\, dx = \frac{1 \cdot 3 \cdot 5 \cdots (2n-1)}{2^{n+1} a^n}\left(\frac{\pi}{a}\right)^{1/2}$

4. $\displaystyle\int_0^\infty x\, \mathbf{e}^{-ax^2}\, dx = \frac{1}{2a}$

5. $\displaystyle\int_0^\infty x^3\, \mathbf{e}^{-ax^2}\, dx = \frac{1}{2a^2}$

6. $\displaystyle\int_0^\infty x^{2n+1}\, \mathbf{e}^{-ax^2}\, dx = \frac{n!}{2}\left(\frac{1}{a^{n+1}}\right)$

7. $\displaystyle\int_0^\infty x^n\, \mathbf{e}^{-ax}\, dx = \frac{n!}{a^{n+1}}$

8. $\int_{-\infty}^{\infty} x^{2n}\, \mathbf{e}^{-ax^2}\, dx = 2\int_{0}^{\infty} x^{2n}\, \mathbf{e}^{-ax^2}\, dx$

9. $\int_{-\infty}^{\infty} x^{2n+1}\, \mathbf{e}^{-ax^2}\, dx = 0$

III TRANSFORMATION OF ∇^2 TO SPHERICAL POLAR COORDINATES

The Laplacian operator has its simplest form in Cartesian coordinates

$$\nabla^2 = \frac{\partial^2}{\partial x^2} + \frac{\partial^2}{\partial y^2} + \frac{\partial^2}{\partial z^2}$$

The transformation and reverse transformation equations from Cartesian to spherical polar coordinates are

$$x = r\sin\theta\cos\phi \qquad r = (x^2 + y^2 + z^2)^{1/2}$$

$$y = r\sin\theta\sin\phi \qquad \cos\theta = \frac{z}{(x^2 + y^2 + z^2)^{1/2}}$$

$$z = r\cos\theta \qquad \tan\phi = \frac{y}{x}$$

We now must determine the transformation derivatives:

$$\frac{\partial r}{\partial x} = \frac{1}{2}(x^2 + y^2 + z^2)^{-1/2} 2x = \frac{x}{r} = \sin\theta\cos\phi$$

Likewise,

$$\frac{\partial r}{\partial y} = \sin\theta\sin\phi; \quad \frac{\partial r}{\partial z} = \cos\theta$$

A simple way to find $\partial\theta/\partial x$ without having to differentiate the inverse cosine is to differentiate $\cos\theta$ directly.

$$-\sin\theta\, d\theta = -z\left(\frac{1}{2}\right)(x^2 + y^2 + z^2)^{-3/2}(2x)\, dx = -\frac{zx}{r^3}\, dx$$

$$-d\theta = -\frac{\cos\phi\cos\theta}{r}\, dx$$

$$\frac{\partial\theta}{\partial x} = \frac{\cos\phi\cos\theta}{r}$$

By the same method, we have

$$\frac{\partial\theta}{\partial y} = \frac{\sin\phi\cos\theta}{r} \quad \text{and} \quad \frac{\partial\theta}{\partial z} = \frac{-\sin\theta}{r}$$

To find $\partial\phi/\partial x$ we differentiate the tangent ϕ directly.

$$\sec^2\phi\, d\phi = -\frac{y}{x^2}\, dx$$

$$\frac{d\phi}{\cos^2\phi} = -\frac{r\sin\theta\sin\phi\, dx}{r^2\sin^2\theta\cos^2\phi}$$

$$\frac{\partial\phi}{\partial x} = -\frac{\sin\phi}{r\sin\theta}$$

Likewise,

$$\frac{\partial\phi}{\partial y} = \frac{\cos\phi}{r\sin\theta} \quad \text{and} \quad \frac{\partial\phi}{\partial z} = 0$$

The transformation equations for the first derivatives are found using the chain rule.

$$\frac{\partial}{\partial x} = \frac{\partial r}{\partial x}\frac{\partial}{\partial r} + \frac{\partial\theta}{\partial x}\frac{\partial}{\partial\theta} + \frac{\partial\phi}{\partial x}\frac{\partial}{\partial\phi}$$

$$\frac{\partial}{\partial x} = \sin\theta\cos\phi\frac{\partial}{\partial r} + \frac{\cos\phi\cos\theta}{r}\frac{\partial}{\partial\theta} - \frac{\sin\phi}{r\sin\theta}\frac{\partial}{\partial\phi}$$

$$\frac{\partial}{\partial y} = \frac{\partial r}{\partial y}\frac{\partial}{\partial r} + \frac{\partial\theta}{\partial y}\frac{\partial}{\partial\theta} + \frac{\partial\phi}{\partial y}\frac{\partial}{\partial\phi}$$

$$\frac{\partial}{\partial y} = \sin\theta\sin\phi\frac{\partial}{\partial r} + \frac{\sin\phi\cos\theta}{r}\frac{\partial}{\partial\theta} + \frac{\cos\phi}{r\sin\theta}\frac{\partial}{\partial\phi}$$

$$\frac{\partial}{\partial z} = \frac{\partial r}{\partial z}\frac{\partial}{\partial r} + \frac{\partial\theta}{\partial z}\frac{\partial}{\partial\theta} + \frac{\partial\phi}{\partial z}\frac{\partial}{\partial\phi}$$

$$\frac{\partial}{\partial z} = \cos\theta\frac{\partial}{\partial r} - \frac{\sin\theta}{r}\frac{\partial}{\partial\theta}$$

To find the second derivatives of these operators, we now must operate each operator on itself. Remember that when $\sin\theta\cos\phi\,\partial/\partial r$ operates on $(\cos\phi\cos\theta/r)\,\partial/\partial\theta$, only the $\cos\phi\cos\theta$ passes through the $\partial/\partial r$ operator. The term $(1/r)(\partial/\partial\theta)$ must be differentiated as a product.

$$\begin{aligned}
\frac{\partial}{\partial x}\frac{\partial}{\partial x} = {} & \sin^2\theta\cos^2\phi\frac{\partial^2}{\partial r^2} + \frac{\sin\theta\cos\theta\cos^2\phi}{r}\frac{\partial^2}{\partial r\,\partial\theta} - \frac{\sin\theta\cos\theta\cos^2\phi}{r^2}\frac{\partial}{\partial\theta} \\
& - \frac{\sin\phi\cos\phi}{r}\frac{\partial^2}{\partial r\,\partial\phi} + \frac{\sin\phi\cos\phi}{r^2}\frac{\partial}{\partial\phi} + \frac{\cos\theta\cos^2\phi\sin\theta}{r}\frac{\partial^2}{\partial\theta\,\partial r} \\
& + \frac{\cos^2\phi\cos^2\theta}{r}\frac{\partial}{\partial r} + \frac{\cos^2\phi\cos^2\theta}{r^2}\frac{\partial^2}{\partial\theta^2} - \frac{\sin\theta\cos\theta\cos^2\phi}{r^2}\frac{\partial}{\partial\theta} \\
& - \frac{\sin\phi\cos\phi\cos\theta}{r^2\sin\theta}\frac{\partial^2}{\partial\theta\,\partial\phi} + \frac{\sin\phi\cos\phi\cos^2\theta}{r^2\sin^2\theta}\frac{\partial}{\partial\phi} - \frac{\sin\phi\cos\phi}{r}\frac{\partial^2}{\partial\phi\,\partial r} \\
& + \frac{\sin^2\phi}{r}\frac{\partial}{\partial r} - \frac{\sin\phi\cos\phi\cos\theta}{r^2\sin\theta}\frac{\partial^2}{\partial\phi\,\partial r} + \frac{\sin^2\phi\cos\theta}{r^2\sin\theta}\frac{\partial}{\partial\theta} \\
& + \frac{\sin^2\phi}{r^2\sin^2\theta}\frac{\partial^2}{\partial\phi^2} + \frac{\sin\phi\cos\phi}{r^2\sin^2\theta}\frac{\partial}{\partial\phi}
\end{aligned}$$

$$\frac{\partial}{\partial y}\frac{\partial}{\partial y} = \sin^2\theta\sin^2\phi\frac{\partial^2}{\partial r^2} + \frac{\sin\theta\cos\theta\sin^2\phi}{r}\frac{\partial^2}{\partial r\,\partial\theta} - \frac{\sin\theta\cos\theta\sin^2\phi}{r^2}\frac{\partial}{\partial\theta}$$
$$+ \frac{\sin\phi\cos\phi}{r}\frac{\partial^2}{\partial r\,\partial\phi} - \frac{\sin\phi\cos\phi}{r^2}\frac{\partial}{\partial\phi} + \frac{\cos\theta\sin^2\phi\sin\theta}{r}\frac{\partial^2}{\partial\theta\,\partial r}$$
$$+ \frac{\sin^2\phi\cos^2\theta}{r}\frac{\partial}{\partial r} + \frac{\sin^2\phi\cos^2\theta}{r^2}\frac{\partial^2}{\partial\theta^2} - \frac{\sin\theta\cos\theta\sin^2\phi}{r^2}\frac{\partial}{\partial\theta}$$
$$+ \frac{\sin\phi\cos\phi\cos\theta}{r^2\sin\theta}\frac{\partial^2}{\partial\theta\,\partial\phi} - \frac{\sin\phi\cos\phi\cos^2\theta}{r^2\sin^2\theta}\frac{\partial}{\partial\phi} + \frac{\sin\phi\cos\phi}{r}\frac{\partial^2}{\partial\phi\,\partial r}$$
$$+ \frac{\cos^2\phi}{r}\frac{\partial}{\partial r} + \frac{\sin\phi\cos\phi\cos\theta}{r^2\sin\theta}\frac{\partial^2}{\partial\phi\,\partial r} + \frac{\cos^2\phi\cos\theta}{r^2\sin\theta}\frac{\partial}{\partial\theta}$$
$$+ \frac{\cos^2\phi}{r^2\sin^2\theta}\frac{\partial^2}{\partial\phi^2} - \frac{\sin\phi\cos\phi}{r^2\sin^2\theta}\frac{\partial}{\partial\phi}$$

$$\frac{\partial}{\partial z}\frac{\partial}{\partial z} = \cos^2\theta\frac{\partial^2}{\partial r^2} - \frac{\sin\theta\cos\theta}{r}\frac{\partial^2}{\partial r\,\partial\theta} + \frac{\sin\theta\cos\theta}{r^2}\frac{\partial}{\partial\theta} - \frac{\sin\theta\cos\theta}{r}\frac{\partial^2}{\partial r\,\partial\theta}$$
$$+ \frac{\sin^2\theta}{r}\frac{\partial}{\partial r} + \frac{\sin^2\theta}{r^2}\frac{\partial^2}{\partial\theta^2} + \frac{\sin\theta\cos\theta}{r^2}\frac{\partial}{\partial\theta}$$

Adding the second derivatives,

$$\frac{\partial^2}{\partial x^2} + \frac{\partial^2}{\partial y^2} + \frac{\partial^2}{\partial z^2} = \frac{\partial^2}{\partial r^2} + \frac{2}{r}\frac{\partial}{\partial r} + \frac{1}{r^2}\frac{\partial^2}{\partial\theta^2} + \frac{\cos\theta}{r^2\sin\theta}\frac{\partial}{\partial\theta} + \frac{1}{r^2\sin^2\theta}\frac{\partial^2}{\partial\phi^2}$$

or, as it usually is expressed,

$$\frac{\partial^2}{\partial x^2} + \frac{\partial^2}{\partial y^2} + \frac{\partial^2}{\partial z^2} = \frac{1}{r^2}\frac{\partial}{\partial r}\left(r^2\frac{\partial}{\partial r}\right) + \frac{1}{r^2\sin\theta}\frac{\partial}{\partial\theta}\left(\sin\theta\frac{\partial}{\partial\theta}\right) + \frac{1}{r^2\sin^2\theta}\frac{\partial^2}{\partial\phi^2}$$

IV STIRLING'S APPROXIMATION

Throughout many areas of physical chemistry, and particularly in the area of statistical mechanics, factorials are used extensively. One will recall that N factorial, written $N!$, is defined as

$$N! = (1)(2)(3)\cdots(N-1)(N) \tag{IV-1}$$

where $0! = 1$ and $1! = 1$.

Let us consider, now, an expression for the natural logarithm of $N!$.

$$\ln N! = \ln 1 + \ln 2 + \ln 3 + \cdots + \ln N$$
$$\ln N! = \sum_{x=1}^{N} \ln x \tag{IV-2}$$

If N is very large, however, this summation can be replaced by the integral

$$\ln N! \cong \int_1^N \ln x\, dx \tag{IV-3}$$

which, when integrated by parts, gives

$$\ln N! \cong x \ln x - x \Big|_1^N = N \ln N - N + 1 \tag{IV-4}$$

However, if N is much larger than 1, we can write

$$\ln N! \cong N \ln N - N \quad \text{for large } N \tag{IV-5}$$

Equation (IV-5) is known as *Stirling's approximation* and can be used to approximate ln $N!$ when N is a very large number.

A much more accurate expression for approximating $N!$, particularly for nonintegral values of N, can be found using gamma functions. This approach, which is beyond the scope of this text, leads to *Stirling's formula*

$$N! = \sqrt{2\pi N}\left(\frac{N}{e}\right)^N \tag{IV-6}$$

which again is valid as N becomes very large. Taking the logarithm of this equation gives

$$\ln N! \cong N \ln N - N + \frac{1}{2} \ln (2\pi N) \quad \text{for large } N \tag{IV-7}$$

Equation (IV-6) gives better results for smaller values of N.

Let us compare the abilities of Equations (IV-4), (IV-5), and (IV-6) to approximate a relatively small number for N: 60!. The actual value is 8.32×10^{81}. Equation (IV-4) gives a value for $60! = 1.16 \times 10^{81}$. Equation (IV-5) gives an even poorer approximation for $60! = 4.27 \times 10^{80}$. Equation (IV-6) gives a value for $60! = 8.31 \times 10^{81}$. Clearly, for an N value as small as 60, Equation (IV-6) gives the best results. As N gets much larger, however, all three equations work better and better. In chemical statistics, where N usually represents the number of molecules in a macroscopic system, 10^{23}, all equations reduce to Equation (IV-5).

Answers

CHAPTER 1

2. (a) $r = 2\sqrt{2},\ \theta = 45^\circ$
(b) $r = \sqrt{3},\ \theta = 54.74^\circ$
(c) $r = 5.10,\ \theta = 78.69^\circ$
(d) $r = 4.12,\ \theta = 345.96^\circ$
(e) $r = 3.61,\ \theta = 213.69^\circ$
(f) $r = 2.00,\ \theta = 0^\circ$
(g) $r = 2.00,\ \theta = 180^\circ$
(h) $r = 5.00,\ \theta = 270^\circ$
(i) $r = 13.42,\ \theta = 333.43^\circ$

3. (a) $x = 0.647,\ y = 0.902$
(b) $x = 1.00,\ y = 0$
(c) $x = -2.23,\ y = -2.23$
(d) $x = 0,\ y = \sqrt{3}$
(e) $x = -4.91,\ y = 3.44$
(f) $x = 0,\ y = -2.50$
(g) $x = 2.46,\ y = 1.72$
(h) $x = 1.58,\ y = 4.74$

4. (a) $r = \sqrt{3},\ \theta = 54.74^\circ,\ \phi = 45^\circ$
(b) $r = 3.74,\ \theta = 74.50^\circ,\ \phi = 33.69^\circ$
(c) $r = 2.24,\ \theta = 116.57^\circ,\ \phi = 0^\circ$
(d) $r = 4.12,\ \theta = 14.04^\circ,\ \phi = 180^\circ$
(e) $r = 13.75,\ \theta = 150.78^\circ,\ \phi = 243.43^\circ$
(f) $r = 4.00,\ \theta = 180^\circ,\ \phi =$ undefined

5. $d\tau = r\, d\theta\, dr\, dz$

6. (a) $|z| = 3,\ \theta = 0^\circ$
(b) $|z| = 6,\ \theta = 90^\circ$
(c) $|z| = 2\sqrt{2},\ \theta = 45^\circ$
(d) $|z| = 3.16,\ \theta = 288.43^\circ$
(e) $|z| = 4\sqrt{2},\ \theta = 225^\circ$
(f) $|z| = 6.40,\ \theta = 128.68^\circ$

9. $0,\ \pm 1,\ \pm 2, \ldots$

CHAPTER 2

1. (a) $x = 1$
(b) $x = 0.5$
(c) $x = 4, -2$
(d) $x = 1, -0.25$
(e) $x = \sqrt{3}, \sqrt{3}$
(f) $x = n\pi, n = 0, \pm 1, \pm 2, \ldots$
(g) $\theta = (2n+1)\dfrac{\pi}{2}, n = 0, \pm 1, \pm 2, \ldots$
(h) $(H^+) = 1.00M$
(i) $x = \pm 2$
(j) none

6. (a) $-\dfrac{1}{x(x+h)}$
(b) $\dfrac{-2x - h}{x^2(x^2 + 2xh + h^2)}$
(c) $4(2x + h)$
(d) $\dfrac{-1}{(1 + x + h)(1 + x)}$

7. $C_0 = 2.30, k = 0.0138$ time^{-1}

8. (a) $x = 1.247, -1.802, -0.445$
(b) $x = \pm 0.618, \pm 1.618$
(c) $x = 1.425, 0.053, -1.000, -1.479$
(d) $x = \pm 1, \pm 1, \pm 2$

CHAPTER 3

1. (a) 7.000
(b) 0.955
(c) 8.025
(d) −0.151
(e) 1.264
(f) −1.079

2. (a) $(H^+) = 1.000M$
(b) $(H^+) = 3.573 \times 10^{-3}M$
(c) $(H^+) = 1.279 \times 10^{-6}M$
(d) $2.786 \times 10^{-8}M$
(e) $5.916M$
(f) $7.674 \times 10^{-13}M$

3. pH = 6.98
4. work = −3003 J
5. $\Delta S = 12.70$ J/K
6. $T_2 = 103$ K
7. $t = 3300$ yr

CHAPTER 4

1. (a) $12x^2 + 14x - 10$

   (b) $\dfrac{-x}{\sqrt{1-x^2}}$

   (c) $4x - 9$

   (d) $6\sec^2 2\theta$

   (e) $2x^3e^{2x} + 3x^2e^{2x}$

   (f) $A[\cos^2\theta - \sin^2\theta]$

   (g) $\dfrac{-x^4e^x}{2\sqrt{1-e^x}} + 4x^3\sqrt{1-e^x}$

   (h) $4x^6(1-e^x)\cos 4x - x^6e^x\sin 4x + 6x^5(1-e^x)\sin 4x$

   (i) $\dfrac{3x^3}{2(1-3x)^{3/2}} + \dfrac{3x^2}{(1-3x)^{1/2}}$

   (j) $-\dfrac{e^x}{(1-e^x)}$

   (k) $-1 - \ln n_i$

   (l) $-3\ln te^{-3t} + \dfrac{1}{t}e^{-3t}$

   (m) $-\dfrac{A}{t^2} + 1 + \ln t$

   (n) $\dfrac{E^2}{A}\left(2z - \dfrac{27}{8}\right)$

   (o) $2A\left(\dfrac{N\pi}{L}\right)\cos\left(\dfrac{N\pi x}{L}\right)$

   (p) $\dfrac{\Delta H}{Rt^2}$

   (q) $\dfrac{\Delta G}{Rt^2}$

   (r) $-\dfrac{12A}{r^{13}} + \dfrac{6B}{r^7}$

   (s) $-\dfrac{M}{v^2}$

   (t) $A\,e^{-B/Rt}\left(\dfrac{B}{Rt^2}\right)$

2. (a) $\left(\dfrac{\partial P}{\partial V}\right)_T = -\dfrac{nRT}{V^2}$

   (b) $\left(\dfrac{\partial P}{\partial V}\right)_T = -\dfrac{nRT}{(V-nb)^2} + \dfrac{2n^2a}{V^3}$

(c) $\left(\frac{\partial \rho}{\partial T}\right) = -\frac{PM}{RT^2}$

(d) $\left(\frac{\partial H}{\partial T}\right) = b + 2cT - \frac{d}{T^2}$

(e) $\left(\frac{\partial r}{\partial z}\right) = \frac{z}{(x^2 + y^2 + z^2)^{1/2}}$

(f) $\left(\frac{\partial y}{\partial \phi}\right) = -r \sin\theta \sin\phi$

(g) $\frac{\partial^2 S}{\partial P\, \partial T} = \frac{1}{T}\left(\frac{\partial^2 H}{\partial P\, \partial T}\right)$

(h) $\frac{\partial^2 S}{\partial T\, \partial P} = \frac{1}{T}\left(\frac{\partial^2 H}{\partial T\, \partial P}\right) - \frac{1}{T^2}\left(\frac{\partial H}{\partial P}\right)_T + \frac{V}{T^2} - \frac{1}{T}\left(\frac{\partial V}{\partial T}\right)_P$

(i) $\frac{\partial D}{\partial \theta} = \cos\phi[\cos^2\theta - \sin^2\theta]$

(j) $\left(\frac{\partial E}{\partial c_A}\right) = \frac{(2c_A H_{AA} + 2c_B H_{AB}) - E(2c_A + 2c_B S_{AB})}{c_A^2 + c_B^2 + 2c_A c_B S_{AB}}$

(k) $\frac{\partial q}{\partial E_i} = -\frac{1}{kT}\sum e^{-E_i/kT}$

(l) $\frac{\partial q}{\partial T} = \frac{1}{kT^2}\sum E_i e^{-E_i/kT}$

3. (a) slope = 6
(b) slope = 25
(c) slope = 4
(d) slope = 2.609
(e) slope = 0
(f) slope = −10
(g) slope = 1.5
(h) slope = 196 m/s
(i) slope = 31.2×10^{-3}
(j) slope = 0.0412
(k) slope = -3.94×10^{-4}

4. (a) minimum at (0.625, 2.438)
(b) maximum at (−3, 97)
minimum at (2, −28)
point of inflection at (−0.5, 34.5)
(c) maxima at $x = (2n+1)\frac{\pi}{6}$, $y = \sin(2n+1)\frac{\pi}{2}$, $n = 0, 2, 4, 6, \ldots$

minima at $x = (2n+1)\frac{\pi}{6}$, $y = \sin(2n+1)\frac{\pi}{2}$, $n = 1, 3, 5, 7, \ldots$

(d) no minimum or maximum
(e) minimum at $r = 2^{1/6}\sigma$, $U(r) = -e$
(f) maximum at $\theta = 19.47^\circ$, $\psi = 2.000$
(g) minimum at $z = 1.688$, $E = -2.848\frac{e^2}{a}$
(h) maximum at $E = \frac{1}{2}kT$
(i) maximum at $x = \frac{a}{2}$, $P(x) = \frac{2}{a}$
(j) minimum at $r = \left(\frac{nB}{N_0 A z^2}\right)^{1/n-1}$

5. $\frac{dk}{dT} = \left(\frac{E_a}{RT^2}\right) Ae^{-E_a/RT}$

6. $\frac{d\rho}{dT} = -\frac{PM}{RT^2}$

7. $\left(\frac{\partial P}{\partial T}\right)_V = \frac{nR}{V - nb}$

8. $\left(\frac{\partial P}{\partial V}\right)_T = -\frac{nRT}{(V - nb)^2} + \frac{2n^2 a}{V^3}$

9. $\alpha = \frac{R}{RT + bP}$

10. $dV = -\frac{RT}{P^2} dP + \frac{R}{P} dT$

11. $r = r_0$

12. $x = a/2$

CHAPTER 5

1. (a) $f(x) = \frac{4}{3}x^3 + C$
   (b) $f(x) = -1/x + C$
   (c) $f(x) = -\frac{1}{3}\cos 3x + C$
   (d) $f(x) = \frac{9}{4}x^4 + 10x^3 + \frac{25}{2}x^2 + C$
   (e) $f(x) = 4e^x + C$
   (f) $f(v) = Pv + C$
   (g) $f(p) = RT \ln p + C$
   (h) $f(v) = \frac{1}{2}Mv^2 + C$
   (i) $f(r) = -\frac{Q^2}{r} + C$
   (j) $f(t) = \frac{1}{2\pi W}\sin(2\pi Wt) + C$

2. (a) $y = -\frac{1}{4}e^{-4x} + C$
   (b) $y = \frac{1}{3}x^3 - A^2 x + C$
   (c) $y = \frac{1}{2}\left[x\sqrt{(x^2 - A^2)} - A^2 \ln\left|x + \sqrt{(x^2 - A^2)}\right|\right] + C$
   (d) $y = \frac{1}{8}x^8 - \frac{1}{3}x^6 + x^4 + C$
   (e) $y = \frac{x}{2} - \frac{A}{4\pi N}\sin\left(\frac{2N\pi x}{A}\right)$
   (f) $y = \frac{x^2}{4} - \frac{A}{4\pi N}x\sin\left(\frac{2N\pi x}{A}\right) - \frac{A^2}{8N^2\pi^2}\cos\left(\frac{2N\pi x}{A}\right) + C$
   (g) $y = \frac{\Delta H}{Rt} + C$
   (h) $y = \frac{1}{2}e^x(\cos x + \sin x) + C$
   (i) $y = \frac{t}{2} - \frac{1}{8\pi W}\sin(4\pi Wt) + C$

(j) $y = -\frac{1}{4}\cos^4\phi + C$

(k) $y = \frac{3\theta}{8} + \frac{3\sin 2\theta}{16} + \frac{\cos^3\theta\sin\theta}{4} + C$

(l) $y = -\frac{1}{18}\sin^5(3x+4)\cos(3x+4) + \frac{5x}{16} - \frac{5}{96}\sin 2(3x+4)$
$- \frac{5}{72}\sin^3(3x+4)\cos(3x+4) + C$

(m) $y = \frac{x^2}{2}\sin 2x - \frac{1}{4}\sin 2x + \frac{x}{2}\cos 2x + C$

(n) $y = \ln\frac{(4-x)}{(3-x)} + C$

(o) $y = \frac{-\Delta H}{t} + A\ln t + \frac{B}{2}t + \frac{C}{6}t^2 + C'$

(p) $y = C_p \ln t + C$

(q) $y = \left[\frac{1}{(n-1)(a-x)^{n-1}}\right] + C$

(r) $y = -\frac{1}{a^2}e^{-ar}(ar+1) + C$

(s) $-kTe^{-\varepsilon/kT} + C$

(t) $\ln(A) = -kt + C$

3. (a) $a(T_2 - T_1) + \frac{b}{2}(T_2^2 - T_1^2) + \frac{c}{3}(T_2^3 - T_1^3) + d\ln\frac{T_2}{T_1}$

(b) $RT\ln\frac{P_2}{P_1}$

(c) 2π

(d) $-\frac{\Delta H}{R}\left(\frac{1}{T_2} - \frac{1}{T_1}\right)$

(e) $nRT\ln\frac{(V_2 - nb)}{(V_1 - nb)} + n^2a\left(\frac{1}{V_2} - \frac{1}{V_1}\right)$

(f) $\frac{1}{3}$

(g) $\frac{a^3}{6} - \frac{a^3}{4n^2\pi^2}$

(h) $\frac{1}{4a}\left(\frac{\pi}{a}\right)^{1/2}$

(i) $\frac{a_0^2}{4}$

(j) $\frac{2k^2T^2}{m^2}$

(k) $1/a$

5. (a) $\frac{1}{6}x^3y^2 + C$

(b) $\frac{1}{3}(x^3y + xy^3) + C$

(c) $\frac{1}{2}y^2(x\ln x - x) + C$

(d) $\left[\frac{x^2e^{2x}}{2} - \frac{e^{2x}}{4}(2x-1)\right]\left(y\ln y - y\right)z + C$

(e) 2

(f) $\frac{4}{3}\pi r^3$

(g) $\frac{abc}{h^3}(2\pi mkT)^{3/2}$

7. $\Delta H = aT + \frac{b}{2}T^2 + \frac{c}{3}T^3 + d$

8. $\Delta G = a - bT\ln T - cT^2 + eT$

9. 0.198

10. 0.0063

11. $\frac{3}{2}a_0$

CHAPTER 6

1. (a) $y = Ae^{-3x}$

(b) $y = Ae^{3x}$

(c) $y = c_1\,e^{-x} + c_2\,x\,e^{-x}$

(d) $y = c_1\,e^{3x} + c_2\,x\,e^{3x}$

(e) $y = A\,e^{i3x} + B\,e^{-i3x}$

(f) $-\frac{1}{k_1+k_2}\ln[k_1 a - (k_1+k_2)x] = t + C$

(g) $\phi = Ae^{-ar}$

(h) $\ln(A) = -kt + C$

(i) $\Phi = A\,e^{im\phi} + B\,e^{-im\phi}$

(j) $y = A\sin\sqrt{\frac{k}{m}}t + B\cos\sqrt{\frac{k}{m}}t$

(k) $\psi = A\sin\sqrt{\frac{8\pi^2mE}{h^2}}x + B\cos\sqrt{\frac{8\pi^2mE}{h^2}}x$

2. (a) Exact

(b) Exact

(c) Exact

(d) Exact

(e) Exact

(f) Inexact

(g) Exact

(h) Exact

(i) Inexact

(j) Exact

(k) Exact

5. $(\kappa^2 - c^2)a_\kappa = 0$

7. $f(x) = A\sin\frac{2\pi x}{\lambda} + B\cos\frac{2\pi x}{\lambda}$

8. $\lambda = \frac{2L}{n}$; $n = 1, 2, 3, 4, \ldots$

CHAPTER 7

1. (a) divergent

(b) divergent

(c) convergent

(d) convergent

(e) convergent

(f) divergent

(g) divergent

(h) convergent

(i) divergent

(j) convergent

2. (a) convergent
(b) divergent
(c) test fails
(d) divergent
(e) convergent
(f) convergent
(g) divergent
(h) convergent
(i) divergent
(j) test fails

3. (a) $-1 < x < 1$
(b) $-1 < x < 1$
(c) all values of x
(d) $-1 < x \leq 1$
(e) all values of x
(f) $-1 \leq x \leq 1$
(g) all values of x
(h) $0 < x \leq 2$
(i) $-2 < x < 2$
(j) $-3 < x < -1$

4. (a) $1 - x + x^2 - x^3 + - \cdots$
(b) $1 + 2x + 3x^2 + 4x^3 + \cdots$
(c) $1 + \frac{1}{2}x - \frac{1}{8}x^2 + \frac{1}{16}x^3 - + \cdots$
(d) $-x - x^2/2 - x^3/3 - x^4/4 - \cdots$
(e) $1 - x^2 + x^4/2! - x^6/3! + - \cdots$
(f) $1 + x \ln a + (x \ln a)^2/2! + (x \ln a)^3/3! + \cdots$
(g) $1 - x^2/2! + x^4/4! - x^6/6! + - \cdots$
(h) $1 + 3x + 3x^2 + x^3$

10. $g(k) = \dfrac{2i}{\sqrt{2\pi}k^2}(k\pi \cos k\pi - \sin k\pi)$

11. $g(k) = \dfrac{\sin kL}{kL}$

CHAPTER 8

1. (a) $|A| = 3.16$, $\theta = 71.57^\circ$
(b) $|A| = 2.83$, $\theta = 45^\circ$
(c) $|A| = 5.00$, $\theta = 306.87^\circ$
(d) $|A| = 2.00$, $\theta = 180^\circ$
(e) $|A| = 6.08$, $\theta = 260.53^\circ$
(f) $|A| = 3.32$, $\theta = 25.37^\circ$, $\phi = 45^\circ$
(g) $|A| = 5.39$, $\theta = 42.08^\circ$, $\phi = 56.32^\circ$
(h) $|A| = 2.45$, $\theta = 114.08^\circ$, $\phi = 116.57^\circ$
(i) $|A| = 3.32$, $\theta = 154.63^\circ$, $\phi = 225^\circ$
(j) $|A| = 1.41$, $\theta = 135^\circ$, $\phi = 0^\circ$

2. (a) $|C| = 5.66$, $\theta = 45^\circ$
(b) $|C| = 4.12$, $\theta = 75.97^\circ$
(c) $|C| = 4.24$, $\theta = 45^\circ$
(d) $|C| = 7.07$, $\theta = 45^\circ$, $\phi = 53.13^\circ$
(e) $|C| = 3.74$, $\theta = 122.33^\circ$, $\phi = 198.43^\circ$
(f) $|C| = 8.54$, $\theta = 134.63^\circ$, $\phi = 99.47^\circ$

3. (a) 6
(b) 2
(c) -4
(d) 9
(e) -34
(f) -33

4. (a) $|C| = 8$, $\theta = 180^\circ$, $\phi =$ undefined
(b) $|C| = 6$, $\theta = 180^\circ$, $\phi =$ undefined

(c) $|C| = 12$
$\theta = 0^\circ$
$\phi = \text{undefined}$

(d) $|C| = \sqrt{6}$
$\theta = 65.91^\circ$
$\phi = 296.57^\circ$

(e) $|C| = 19.52$
$\theta = 55.70^\circ$
$\phi = 262.87^\circ$

(f) $|C| = 23.43$
$\theta = 59.20^\circ$
$\phi = 153.43^\circ$

8. $L_x = y\,p_z - z\,p_y \qquad L_y = z\,p_x - x p_z \qquad L_z = x p_y - y\,p_x$

CHAPTER 9

1. (a) -2
(b) -5
(c) -4
(d) 1
(e) $x^2 - 1$
(f) 1
(g) 18
(h) -93
(i) $x^3 - 2x$
(j) $+352$
(k) $x^4 - 3x^2b^2 + b^4$

2. (a) $x = \pm 1$
(b) $x = \pm 2$
(c) $x = \pm\sqrt{5}$
(d) $x = 0, \pm\sqrt{3}$
(e) $x = 0, 0, \pm\sqrt{3}$

3. $$\begin{pmatrix} 5 & 1 & 0 & 6 \\ 5 & 3 & -6 & 7 \\ -2 & 3 & 3 & -3 \\ 0 & 8 & 9 & 12 \end{pmatrix}$$

4. (a)
$$\begin{pmatrix} -6 & 1 \\ 12 & -7 \end{pmatrix}$$
(b)
$$\begin{pmatrix} 4 & -1 \\ 2 & 3 \end{pmatrix}$$
(c)
$$\begin{pmatrix} 12 & 15 & 18 \\ 3 & -1 & -7 \\ 12 & 23 & 38 \end{pmatrix}$$
(d)
$$\begin{pmatrix} 26 & 14 & -18 \\ 10 & 15 & 7 \\ 18 & -14 & -43 \end{pmatrix}$$
(e)
$$\begin{pmatrix} x + 8y + 4z \\ -2x + 3y \\ 5x - y - z \end{pmatrix}$$

5. $$\begin{pmatrix} 42 & -13 & 11 \\ 68 & -50 & 36 \\ 12 & 6 & -2 \end{pmatrix} \quad \text{vs} \quad \begin{pmatrix} 8 & 0 & 34 \\ 4 & -7 & 37 \\ 12 & 29 & -11 \end{pmatrix}$$

6. (a) $x = 2, y = 1$
(b) $x = 1, y = 3, z = -4$
(c) $x = 1, y = 1, z = -2, t = -3$
(d) $x = x' \sin\theta - y' \cos\theta$
$y = y' \sin\theta + x' \cos\theta$

11. $\mathbf{\Lambda} = \begin{pmatrix} 4 & 0 \\ 0 & -1 \end{pmatrix}$

14. $E_1 = \alpha + 2\beta; \quad E_2 = E_3 = \alpha - \beta$

For E_1: $c_1 = c_2 = c_3 = \dfrac{1}{\sqrt{3}}$

For E_2: $c_1 = \dfrac{1}{\sqrt{2}}, c_2 = -\dfrac{1}{\sqrt{2}}, c_3 = 0$

For E_3: $c_1 = \dfrac{1}{\sqrt{6}} = c_2, c_3 = -\dfrac{2}{\sqrt{6}}$

CHAPTER 10

1. (a) $1+x+x^2+x^3+x^4+x^5$
(b) $1-x+x^2-x^3+x^4-x^5$
(c) $\Delta E = E_2 - E_1$
(d) $3x^2y$
(e) $2y^3$
(f) $12x^3y^2$
(g) $8xyz$
(h) $1+2x+3x^2+4x^3+5x^4$
(i) $x_0! \cdot x_1! \cdot x_2! \cdot x_3! \cdot x_4!$
(j) $\begin{pmatrix} -1 & 0 \\ 0 & -1 \end{pmatrix} \begin{pmatrix} a \\ b \end{pmatrix} = \begin{pmatrix} -a \\ -b \end{pmatrix}$

2. (a) commute
(b) commute
(c) commute
(d) do not commute

7. eigenvalues $= m\hbar$

9. eigenvalues $= a^2$

10. (a) (0.732, 2.732)
(b) (2.121, 3.535)
(c) (4, 3)
(d) (−0.232, 3.598)
(e) (−3.098, 0.634)

11.
$$\hat{M}_x = -i\hbar \left(y\frac{\partial}{\partial z} - z\frac{\partial}{\partial y} \right)$$
$$\hat{M}_y = -i\hbar \left(z\frac{\partial}{\partial x} - x\frac{\partial}{\partial z} \right)$$
$$\hat{M}_z = -i\hbar \left(x\frac{\partial}{\partial y} - y\frac{\partial}{\partial x} \right)$$

12.
$$\hat{M}_x = i\hbar \left(\sin\phi\frac{\partial}{\partial\theta} + \cot\theta\cos\phi\frac{\partial}{\partial\phi} \right)$$
$$\hat{M}_y = i\hbar \left(-\cos\phi\frac{\partial}{\partial\theta} + \cot\theta\sin\phi\frac{\partial}{\partial\phi} \right)$$
$$\hat{M}_z = -i\hbar\frac{\partial}{\partial\phi}$$

13.
$$\hat{M}^2 = -\frac{h^2}{4\pi^2}\left(\frac{\partial^2}{\partial\theta^2} + \cot\theta\frac{\partial}{\partial\theta} + \frac{1}{\sin^2\theta}\frac{\partial^2}{\partial\phi^2} \right)$$

CHAPTER 12

1. (a) $\frac{1}{36}$
(b) $\frac{1}{18}$
(c) $\frac{1}{12}$
(d) $\frac{1}{9}$
(e) $\frac{5}{36}$
(f) $\frac{1}{6}$
(g) $\frac{1}{18}$

2. (a) 5.649^4
(b) ± 0.02
(c) ± 0.001
(d) 0.012
(e) 0.001
(f) 0.014

3. $V = 0.0113 \pm 0.0015$ cc; radius must be measured to a higher precision.

4. $V = 20.2806 \pm 0.0007$ cc

5. $M = 112.58 \pm 0.25$ g/mol

Index

A

Abscissa, definition of, 2
Absolute value, definition of, 8
Antiderivatives, definition of, 54
Antilogarithm, definition of, 26
Areas:
  under curves, 59–61
  graphical determination of, 188
  numerical determination of, 169–174
Arithmetic mean, 182
Associated Legendre function, 83
Associated spherical harmonics, 83
Auxiliary equation, 75
Average error, 184

B

Base of logarithms, 24
Boundary conditions, 92, 139
Boyle's law, graph of, 3

C

Calculus:
  differential, 30–53
  integral, 54–68
Cartesian coordinates, 2–4
  definition of, 2
  differential volume element, 4
  n-dimensional, 3
  three-dimensional, 3
  two-dimensional, 2
Characteristic equation of a matrix, 128–131
Characteristic of a logarithm, 26
Circles:
  equation of, 19
  graph of, 19
Cofactors, method of, 121
Column matrix, 120
Commutation. *See* Commutator
Commutator, 136
Commutator bracket, 136
Comparison test, 95
Complex conjugate, definition of, 8
Complex numbers, definition of, 7
Complex plane, 7–9
  definition of, 8
  graph of, 8
Computer:
  arrays, 160
  loops, 156
  programming, 163–169
  terms:
    `DEF TRUNCATE`, 163
    `DIM`, 160
    `DO-UNTIL`, 158
    `FOR-NEXT`, 156
    `IF-THEN`, 158
    `INPUT-OUTPUT`, 154
    `LPRINT`, 155
    `WHILE (NOT EOF())`, 162
    `WHILE-WEND`, 158
  variables, 152
Constant coefficients, linear differential equations with, 74
Constrained maxima and minima, 47
Convergence:
  of infinite series, 94, 95
  interval of, 99, 101
Convergence and divergence, tests for, 95–99
Convergent series, sums of, 94
Coordinate systems, 1–10
  Cartesian:
    definition of, 2
    differential volume element, 4
    n-dimensional, 3
    three-dimensional, 3
    two-dimensional, 2
  curvilinear, 5
  cylindrical, 10
  plane polar, 4–5
    definition of, 4
    reverse transformation equations, 5
    transformation equations, 5

Coordinate systems (*continued*)
spherical polar, 5–7
definition of, 5
differential volume element, 6
reverse transformation equations, 6
transformation equations, 6
Cosine, definition of, 19
graph of:
linear coordinates, 21
polar coordinates, 21
Cramer's rule, 127
Cross product, 115
Curves, areas under, 61, 169, 188

D

Definite integral. *See* Integrals
∇^2 operator, 138
transformation to plane polar coordinates, 145
transformation to spherical polar coordinates, 206
Derivatives:
definition of, 30
functions of several variables, 35
functions of single variable, 31
geometric properties of, 43–46
mixed partial second, 36
partial, definition of, 36
physical significance of. *See* Derivatives, geometric properties of
relationship to slope, 40, 43
second, 31, 44
Descartes, René, 2
Determinants, definition of, 121
secular, 129
Differential calculus, 30–53
Differential equations, 69–93
definition of, 69
first order, 71
linear, 69
order of, 69
partial, 88
reduced, 70
second order, constant coefficients, 74
series method of solution, 77
special polynomial solutions:
associated Laguerre polynomials, 81
associated Legendre polynomials, 83
Hermite polynomials, 79
Laguerre polynomials, 81
Legendre polynomials, 83
with constant coefficients, 74
general methods of solution, 74
general solutions, 75
imaginary solutions, 75
particular solutions, 75
real solutions, 75
Differential volume element:
Cartesian coordinates, 4
definition of, 4
spherical polar coordinates, 6
Differentials:
exact, definition of, 84
significance of, 86
inexact, definition of, 84
partial, 38
of a cylinder, 39
total, 37–39
Differentiation:
definition of, 31
functions of several variables, 35
functions of single variable, 31
Direct proportion, 13
Divergence:
of infinite series, 94
of a vector, 138
Dot product, 114

E

Eigenfunctions, 138
Eigenvalue equation, 138
Eigenvalue spectrum, 129
Eigenvalues, 128, 138
Eigenvectors, 128, 138
Elements:
of a matrix, 120
of a set, 11
Equations:
auxiliary, 75
differential. *See* Differential equations

eigenvalue. *See* Eigenvalue equations
exponential, 18
defining equation, 18
graph of, 18
zero of, 18
first degree, 12
defining equation, 13
graph of, 13
slope of, 14
zero of, 13
Hermite's, 79
indicial, 78
Laguerre's, 80
Legendre's, 82
linear. *See* First degree equations
logarithmic, 18
defining equation, 18
graph of, 19
polynomial, finding roots of, 20, 174
quadratic. *See* Second degree equations
recursion, 78
second degree, 15
defining equation, 15
graph of, 15
slope of, 15–16
zero of, 15–17
secular, 129
simultaneous. *See* Simultaneous equations
Error:
average, 184
probable, 183
propagation of, 185
random, 182
standard, 183
systematic, 182
Error probability function, 182
Estimated standard deviation, 183
Exact differentials. *See* Differentials, exact
Exactness, test for, 84
Expansion by cofactors, 121
Exponential *e*, definition of, 18
Exponential equations, 18
Exponential functions. *See* Functions, exponential
Euler's relations for complex exponentials, 9, 10
Euler's test for exactness, 84

F

Factorials, 8
First degree equations, 12
defining equations, 13
graph of, 13
slope of, 14
zero of, 13
Formula, recursion, 78
Fourier series, 101–106
Fourier integral, 102
Fourier transforms, 101–106
French curves, 188
Functional dependence of variables, 11, 30
Functions:
circular, 19
graph of, 19
constrained maxima and minima, 47
definition of, 11
differentiation of, 30–53
exponential, 18
graph of, 18
zero of, 18
graphical representation of, 12–20
linear, 12
graph of, 13
slope of, 14
zero of, 13
logarithmic, 18
graph of, 19
maximization of, 43
minimization of, 43
quadratic, 15
graph of, 15
slope of, 15–16
zero of, 15–17
trigonometric, 20
Functions of state, 86

G

Gradient operator, 138
Graphical representation of data, 187

Graphs:
  method of least squares, 165
  preparation of, 187

H

Hamiltonian operator, 140
Heat capacity, definition of, 40
Heisenberg uncertainty principle, 124, 148
Hermite polynomials, 79
Hermitian operators, 140
Heterogeneous logarithms, 26
Homogeneous logarithms, 26

I

Imaginary numbers, 7
Inexact differentials. *See* Differentials, inexact
Infinite series, 94–109
  comparison, 95
  convergence of, 94–99
    tests for, 94–99
  definition of, 94
  divergence of, 94–99
    tests for, 94–99
  Fourier, 101–106
  Maclaurin, 99
  power. *See* Power series
  Taylor, 99
Integral calculus, 54–68
Integrals:
  as an antiderivative, 55
  as area under curve, 59–63
  cyclic, 87
  definite:
    definition of, 61
    evaluation of, 61
    table of, 205
  definition of, 55
  double, 63
  Fourier, 102
  geometric interpretation of, 59–63
  indefinite, table of, 192
  line, 61–63, 86
  partial, 63–64
  triple, 63
Integral sign, 55
Integrand, 55
Integrating factors, 73, 87
Integration:
  definition of, 54
  general methods of, 55
  partial, successive, 64
  special methods of:
    algebraic substitution, 57
    partial fractions, 58
    trigonometric transformation, 58
  as a summation, 59–61
Interval of convergence, 99,101

K

Kronecker delta, 104

L

Lagrange's method of undetermined multipliers, 47
Laguerre polynomials, 81
Laplacian operator, 138
Least squares determination, 165
Legendre polynomials, 83
Line integrals, 61–63, 86
Linear combinations, 70
Linear differential equations. *See* Differential equations, linear
Linear equations, 12
  simultaneous, solutions of, 126
Linear functions. *See* Functions, linear
Linear regression, 165
Logarithmic equations, 18
Logarithmic functions. *See* Functions, logarithmic
Logarithms, 24–29
  base *e*. *See* Logarithms, natural
  base 10. *See* Logarithms, common
  characteristics, 26
  common, 25–26
    relationship to natural, 27
  general properties of, 24–25
    power rule, 25
    product rule, 24
    quotient rule, 25

heterogeneous, 26
homogeneous, 26
mantissas, 26
Napierian. *See* Logarithms, natural
natural, 27–28
relationship to common, 27
significance of, 27–28

M

Maclaurin series, 99
Mantissa, 26
Mathematical sets, 11
Matrices:
addition of, 123
general properties of, 123
multiplication of, 124
putting in diagonal form, 128–131
Matrix:
characteristic equation of, 128
column, 120
definition of, 120
determinant of, 121
diagonal, 125
elements of, 120
inverse, 125
nonsingular, 125
order of, 121
row, 120
singular, 125
square, 121
unit, 125
Matrix algebra, 123–125
Maxima and minima, 43–44
constrained, 47
definition of, 43
Maximization of functions, 43–44
Method of chords, 189
Minimization of functions, 43–44
Modulus. *See* Absolute values, definition of
Monte Carlo method, 172

N

N-factorial:
approximation of, 208
definition of, 8
Newton-Raphson method, 175
Normalization, 104
Numerical methods:
integration:
trapezoid method, 169
Monte Carlo method, 172
Simpson's rule, 171
linear regression, 165
roots to equations:
Newton-Raphson method, 175
Regula Falsi method, 174

O

Operators:
addition of, 140
commutation of, 136
definition of, 135
differential, 135
eigenfunctions of, 138
gradient, 138
Hamiltonian, 140
Hermitian, 140–141
Laplacian, 138
rotational, 142–145
self-adjoint, 140
transformation, 144
vector, 137
Ordinate, definition of, 2
Origin, definition of, 2
Orthogonality:
of functions, 104
of vectors, 115

P

Parabolas, definition of, 15
Partial derivatives. *See* Derivatives, partial
Partial differentials. *See* Differentials, partial
Phase angles, definition of, 8
Physical constants, table of, 192
Plane polar coordinates, 4–5
definition of, 5
graph of, 5
reverse transformation equations, 5
transformation equations, 5
Planimeter, 189

Point of inflection:
  definition of, 44
  test for, 44
Power series:
  definition of, 98
  Fourier, 101
  interval of convergence, 99, 101
  Maclaurin, 99
  Taylor, 99
Probability, definition of, 180
Probability aggregate, 182
Probable error, 183
  of the mean, 184
  of a single measurement, 184
Propagation of errors, 185

Q

Quadratic equations. *See* Second degree equations
Quadratic formula, 16
Quadratic functions. *See* Functions, quadratic

R

Radioactive decay, 23, 29
Random error, 182
Ratio test, 97
Real numbers, 7
Rectangular coordinates. *See* Cartesian coordinates
Recursion equation, 78
Recursion formula, 78
Reduced differential equations, 70
Residuals, 182
Reversibility, 40, 62
Reversible process. *See* Reversibility
Right hand rule, 115–116
Roots. *See under* Zeros
Rotational operators, 142–145
  relationship to symmetry, 144–145
Round space. *See* Space, round
Row matrix, 120

S

Scalar product, 114
Scalars:
  definition of, 110
  transformation properties of, 110
Second degree equations, 15
  defining equations, 15
  graph of, 15
  slope of, 15–16
  zero of, 15–17
Secular determinant, 129
Secular equations, 129
Self-adjoint operators, 140
Separation of variables, 71
Series. *See* Infinite series; Power series
Series expansion of functions. *See* Fourier series; Maclaurin series; Taylor series
Sets. *See* Mathematical sets
Simpson's rule, 171
Simultaneous equations, solutions of, 126
Sine, definition of, 19
Slope:
  determination of using differential calculus, 43
  of linear functions, 14
  of quadratic functions, 15
  variation of (graph), 16
Space:
  rectangular, 2
  round, 4, 5
Spherical polar coordinates, 5–7
  definition of, 5
  differential volume element, 6
  graph of, 6
  reverse transformation equations, 6
  transformation equations, 6
Square matrix, 121
Standard deviation. *See* Standard error
Standard error, 183
  of the mean, 184
State functions, 86
Stirling's approximation, 208
Straight lines, 12
  equations for, 13
  method of least squares. *See* Linear regression
Successive approximations, 163, 174

Successive partial integration, 64
Systematic errors, 182

T

Tangent, definition of, 19
determination of using differential calculus, 43–46
graphical determination of, 188
Taylor series, 101
Total differentials. *See* Differentials, total
Transformation of coordinates, 5, 6, 142–145
Transformation of ∇^2:
to plane polar coordinates, 145–147
to spherical polar coordinates, 206
Trapezoid method of integration, 169
Trigonometric identities, 142

U

Undetermined multipliers, 47
definition of, 47
Unit circle, definition of, 19
Unit vectors, 111
cross product, 116
dot product, 115

V

Variables:
BCD floating point, 152
fractional change, 27
functional dependence of, 11, 30
integer, 152
separation of, 71
string, 152
Vector operators, 137–138
Vector product, 115
Vectors:
absolute value of, 112
addition of, 111–114
applications of, 117–118
definition of, 111
magnitude of, 112
scalar multiplication of, 114
unit, 111
vector multiplication of, 115
Volume elements. *See* Differential volume element
Volumes, by triple integration, 63

W

Work:
area under a curve, 59, 62
pressure-volume, 41, 59, 62, 87
World line, 10

X

x-axis, definition of, 2

Y

y-axis, definition of, 2
y-intercept, 13

Z

Zeros:
of exponential functions, 18
of linear functions, 13
of quadratic functions, 15–16